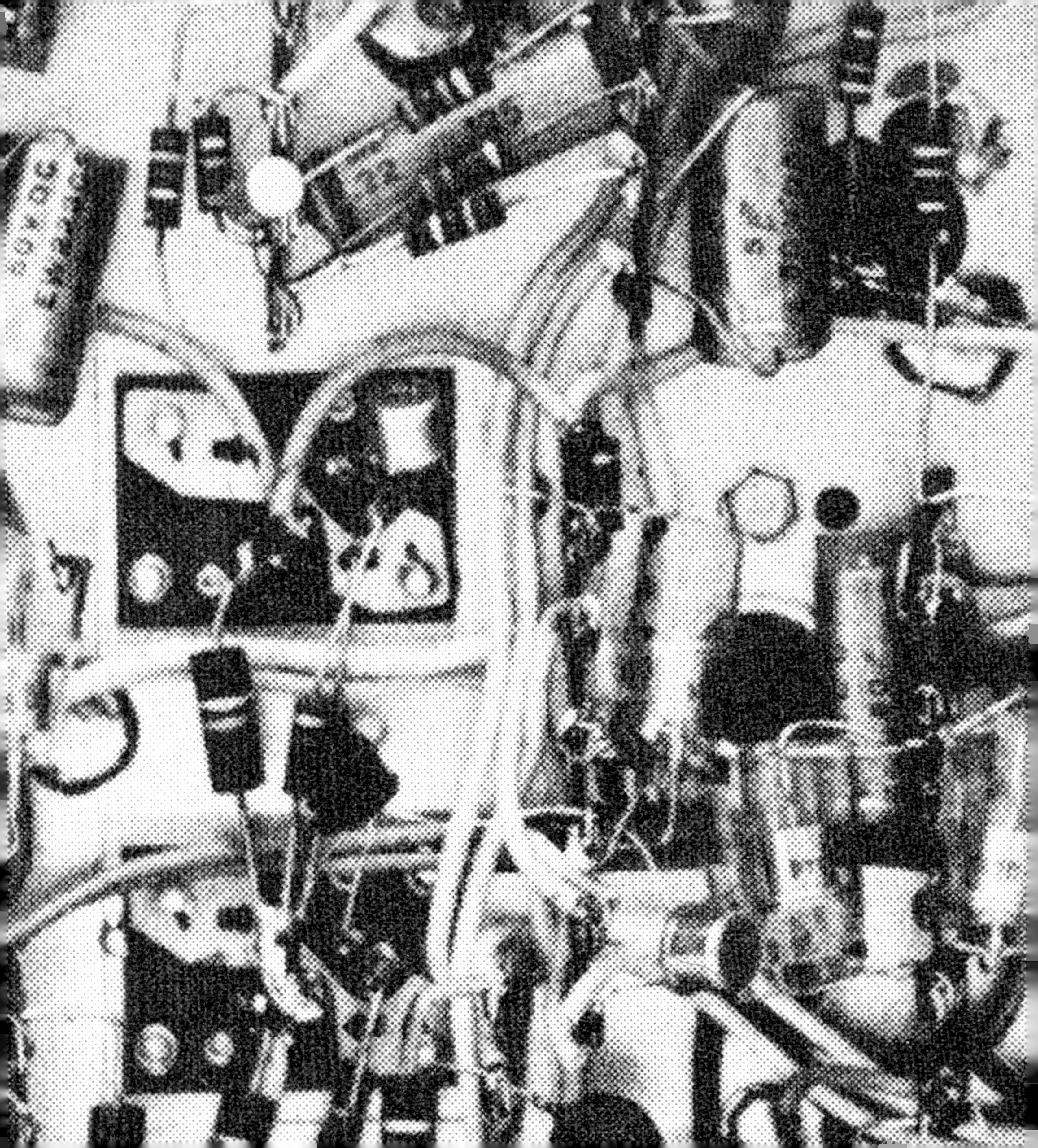

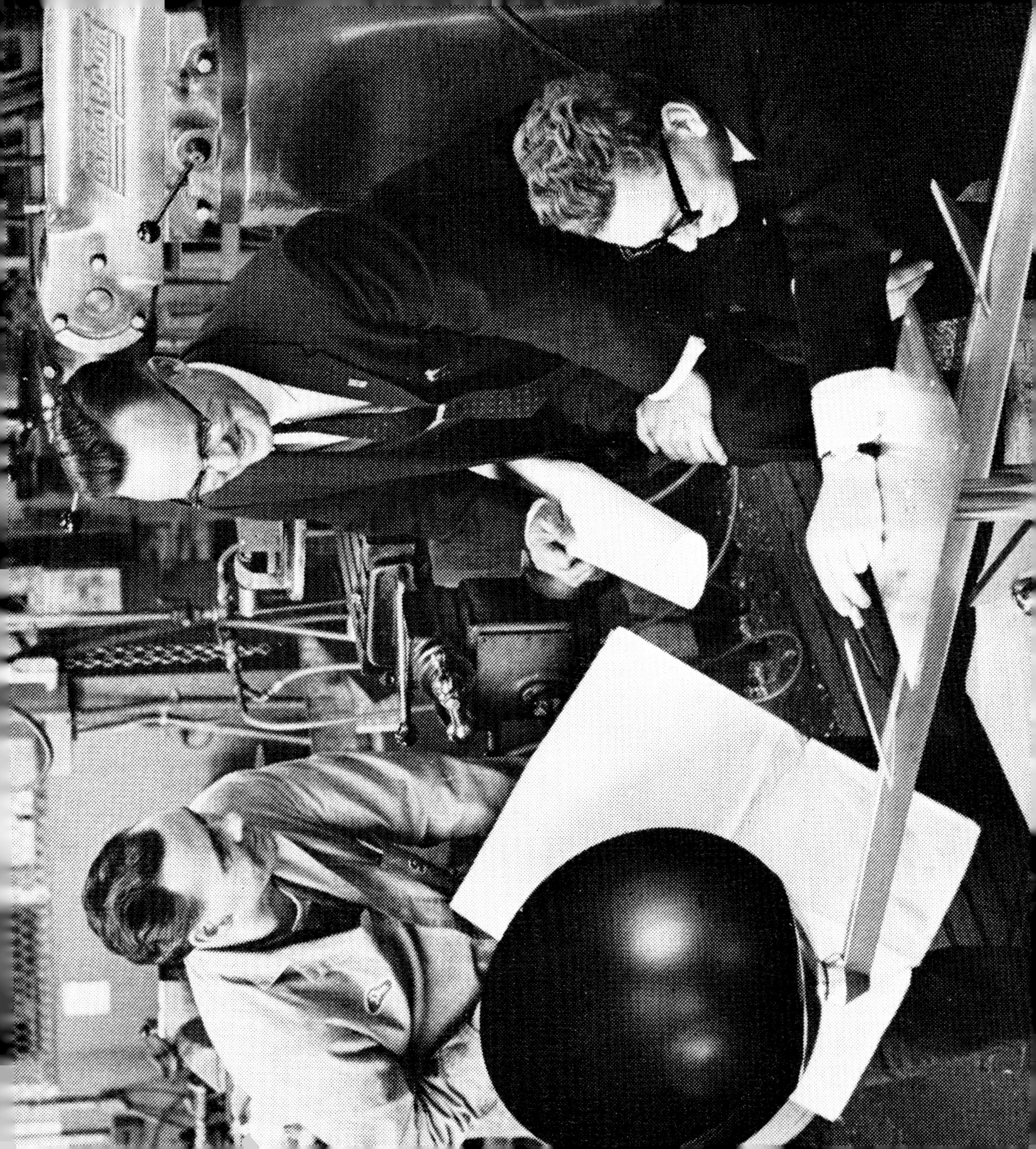

The Art of Clairtone

10 pc 7/32" 84" X 48" I/IV

for Rall cut out

100 pc 9/16" 45-" X 17 1/4 I/III

to be jumed on 4' X 8' car

to 5 pc each

229h.

Neuen Ends for Rame

Nina Munk and Rachel Gotlieb

The Art of Clairtone

The Making of a Design Icon 1958-1971

McClelland & Stewart / Design Exchange
Toronto

Published by McClelland & Stewart

1 2 3 4 5   12 11 10 09 08

Library and Archives Canada Cataloguing in Publication
Munk, Nina
The art of Clairtone / Nina Munk and Rachel Gotlieb.

ISBN 978-0-7710-6507-1 (bound)

1. Clairtone Sound Corporation—History. 2. Electronic industries—Canada—History. I. Gotlieb, Rachel II. Title.

HD9696.A3C343 2008   338.476213810971   C2007-906189-3

Jacket and book design by Barb Woolley and Emese Ungar-Walker, Hambly & Woolley Inc.
Jacket photography by George Whiteside
Pre-press by Somerset Graphics Co. Ltd.
Printed and bound in Canada by Friesens
Printed on acid-free paper

DESIGN EXCHANGE
PROMOTING CANADIAN DESIGN

RIGHT Katyha, Clairtone's lead model and muse, poses on top of a G-TV.

PAGES i-ii Detail of Project G's speaker grill.

PAGE iii Detail from an "Absolutely Shocking" advertisement for Project G.

PAGE 1 "Head in a whirl, discotheque girl Katyha flips to the wild sound of the new G2 Stereo by Clairtone," to quote Clairtone's public relations department.

PAGES 2-3 Frank Sinatra, Davey Davison, and Project G in Marriage on the Rocks (Warner Brothers, 1965).

PAGES 4-5 Detail of the fully transistorized TC4 amplifier chassis, introduced in 1964.

PAGE 6 Designer Hugh Spencer (at right), with staff of Supreme Aluminum Industries, examines a Project G base.

PAGE 8 Project G specs from cabinet-maker George Weil's notebook, 1962.

PAGE 9 David Gilmour and Peter Munk in Clairtone's offices, 1961.

PAGES 12-13 An early Project G with a TC4 chassis and ELAC turntable.

PAGES 174-175 Still from Clairtone's "New York" commercial, 1967.

PAGE 176 Detail of Project G cabinet.

PAGE 177 Control panel of a T10 chassis.

stereo fm
phono
tape
tv
mono
reverse
afc
fm
am
ext spkr
spkrs off
ac on
stereo earphones
PROJECT G
CLAIRTONE
stereo
fm
stereo
balance
VOLUME
BALANCE
BASS
TREBLE
TUNER

Contents

CLAIRTONE
CLAIRTONE
the world's most successful

Foreword
by Samantha Sannella, President, CEO, Design Exchange

Design Exchange is Canada's national centre for the promotion of design. Housed in the historic Toronto Stock Exchange Building, DX creates more than 50 programs each year to teach people about the value of design and its contribution to the economy, to our quality of life, and to the environment.

In 1991, Design Exchange was given a G2 stereo by Frank Davies, one of the many talented designers who worked for Clairtone Sound Corporation during its short history. His donation marked the beginning of DX's collection of Canadian design. In August 1996, DX struck a permanent collection committee to assemble Canadian design. Since then, the collection has grown to include historic and contemporary examples of decorative arts, furniture, graphic design, housewares, lighting, medical equipment, sporting goods, and tableware. But nothing is more distinctive than Clairtone's G series. The Clairtone collection and archive at DX represent a major portion of our holdings and are an important part of Canadian design history. Clairtone represented the best of 1960s design.

To celebrate Clairtone's 50th anniversary, Design Exchange is honoured to work with the publishers McClelland & Stewart, the authors Nina Munk and Rachel Gotlieb, and the designer Barb Woolley to present *The Art of Clairtone: The Making of a Design Icon*.

The "art" in the book's title has multiple meanings—the graphic design, the industrial design of the Project G series, the engineering, and the photography. Perhaps most important is the manner in which all those elements came together to create a powerful brand that stood for style and innovation and optimism.

The creative team behind Clairtone brilliantly used the Clairtone logo, images, messages, and the renowned Project G itself to project a lasting impression of "cool." Clairtone's relationship with celebrities like Frank Sinatra and Oscar Peterson only added to the allure, as did its gregarious founders, Peter Munk and David Gilmour—resplendent in stylish suits, young, and "with-it."

For an object to become an icon, it must trigger a memory so significant that one instantly recognizes and recalls its importance: Tiffany blue, Burberry check, a Saarinen table, an Eames chair, and the Arne Jacobsen Egg chair, used cleverly in Clairtone's marketing campaigns. Project G, which relies on the most simple geometric forms in perfect proportion, inspires and delights with its contrasting materials—perforated metal and rosewood. As in all good design, which balances art and science, the sound of the G is enhanced by the spherical speakers.

Project G is the legacy of Clairtone, the finest product produced during the company's single decade of existence. Conceptually original, technically and functionally perfect, and aesthetically superior, Project G is the epitome of a design icon.

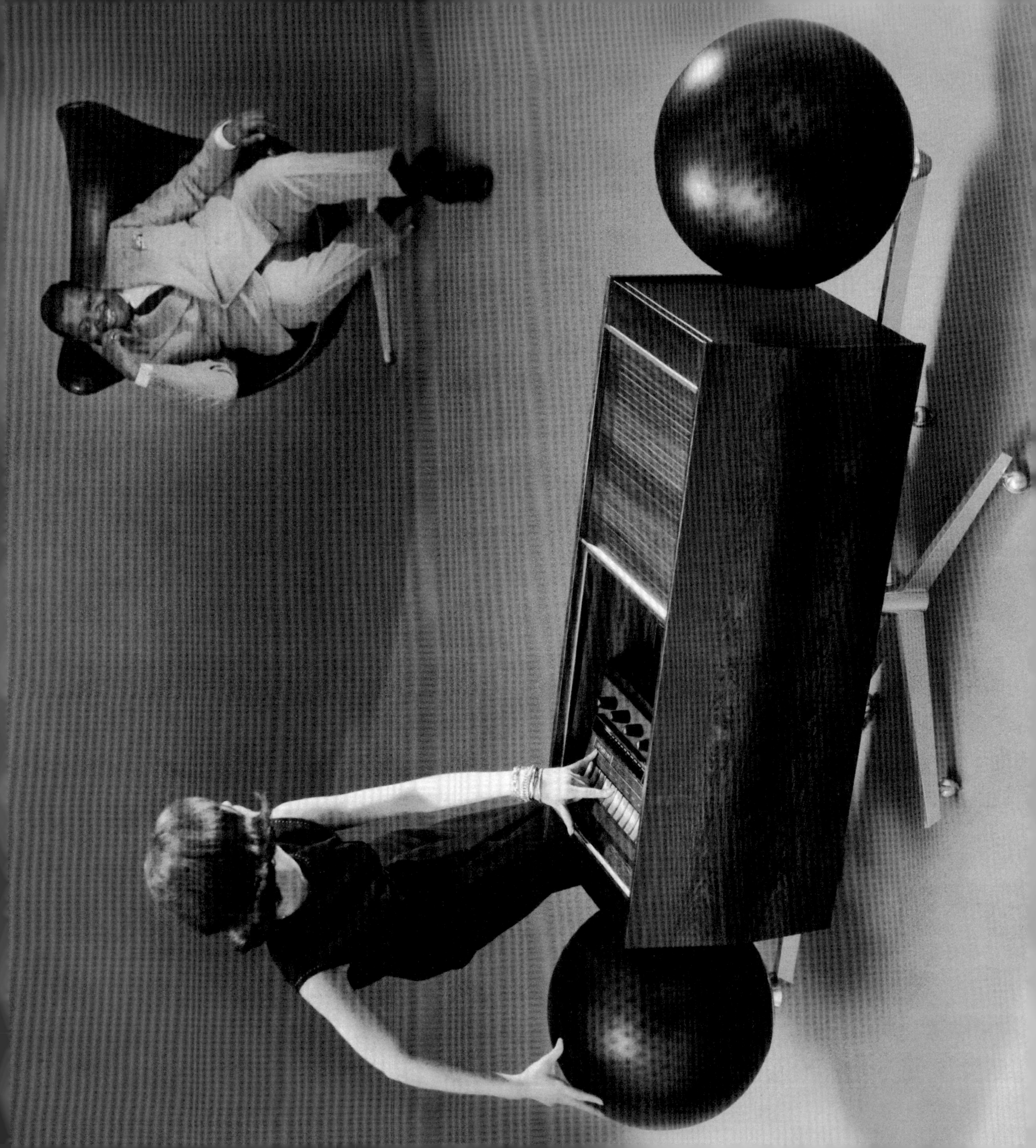

"**Project G epitomized high design, pure form, the perfection of how sacred platonic geometry can bring our banal, everyday products to a higher spiritual art form.** It set the trend for stereo equipment to become a composition of extreme pure form, sophisticated materials, and floating sculptural composition for Bang & Olufsen, Brionvega, Telefunken, and others to emulate."

Karim Rashid

INDUSTRIAL DESIGNER, FOUNDER, KARIM RASHID INC.

In one of Clairtone's most memorable and suggestive marketing images, jazz legend Oscar Peterson is seated in Arne Jacobsen's Egg chair, relaxed and smiling. At the same time, an elegant woman presses a switch on the Project G and strokes one of its "sound globes."

Peter Munk, David Gilmour, and the Memory of Clairtone

by Nina Munk

Clairtone's founders, Peter Munk (left) and David Gilmour, in 1963. "The boys," as they were often referred to back then, were a perfect team.

I WAS born under a falling star, in 1967, the year Clairtone Sound Corporation collapsed. My father remembers it as the worst year of his life. Clairtone was his first company, his "first love," he once called it nostalgically. Measured coldly in dollars and cents, it was his smallest and least successful company; yet nothing my father has done since then has affected him the way Clairtone did.

A few years ago, long after making a name for himself in the gold business, and decades after Clairtone had become little more than a quirky footnote in his career, he confessed to the *New York Times*: "Clairtone was the single most formative experience in my life because it was so traumatic."

But I'm getting ahead of myself. Once upon a time, for a short time, Clairtone Sound Corporation was one of Canada's most dazzling, most admired companies. It started in 1958 with four employees, $3,000, and a cramped, makeshift factory at 26 Sable Street in a Toronto suburb. The initial idea was simple: to merge contemporary Scandinavian furniture design with the latest in high-fidelity equipment.

My father, then 30 and an electrical engineer, made custom hi-fi sets for wealthy clients. His friend and partner, David Harrison Gilmour, 26, had a small business importing Scandinavian flatware, ceramics, and glass. Together, and inspired by a basic 1950s Danish sideboard, they came up with their first hi-fi model—a long, low cabinet in oiled teak with sliding doors and tapered legs. It was good-looking and functional, and it was unlike anything being made in Canada back then. Fitted inside the wooden cabinet were a Dual 1004 turntable, a Granco tube chassis, and a pair of Coral speakers hidden behind plain, wheat-coloured broadcloth from Knoll International.

In 1958, Clairtone's 100-S won a National Design Council award. Peter Munk, left, with other winners: Court Noxon (Metalsmiths), John Stene (Brunswick Manufacturing), and Reinhold Koller (AKA Furniture).

In March 1959, less than four months after it was put into production, that first model, the 100-S, won a Design Award from the National Industrial Design Council. Other models followed, including the entry-level 400-S ("the Princess") and the luxurious 1000-S ("the Signature") with its wireless remote control. Then, almost overnight, it seemed, Clairtone's stereo consoles were everywhere.

"Everybody knew about Clairtone," my father would later boast to the columnist Joan Sutton. "The Prime Minister had one, and if the local truck driver didn't have one, he wanted one." Oscar Peterson, the legendary Canadian jazz pianist, officially endorsed Clairtone. Dizzy Gillespie and Frank Sinatra were avid fans. "Listen to Sinatra on Clairtone stereo. Sinatra does," was one of the company's most memorable tag lines.

During the company's first five years, between 1958 and 1963, production soared from 350 units a year to 25,000 units. The pace was incredible. That year, 1963, Clairtone was listed on the Toronto Stock Exchange. And nothing, nothing at all thrilled my father so much as seeing his upstart company listed alongside Canada's old guard, the Establishment: Massey-Ferguson, Algoma Steel, Canadian Pacific Railway, Abitibi Power & Paper, Falconbridge Nickel, Walker-Gooderham & Worts. "In those days, the TSE was as WASP a club as you can get—it was *it*," my father reminded me. "I was not only not WASP—I was Jewish, I was an immigrant, and I had an accent."

So great was demand for the company's products in the early 1960s that, for a time, at Simpson's department store in Toronto, one Clairtone hi-fi was sold every three hours. In those years, throughout the December rush, Clairtone's factory stayed open around the clock, with cabinetmakers and assembly line workers pushing out stereo consoles in time for Christmas. Keeping up with the orders was exhausting and exhilarating. "I worked 70- and 80-hour weeks," the company's former comptroller, David Pols, told me with pride, echoing other Clairtone employees I interviewed. "Sometimes, I recall, I worked all night."

Remarkably, back in the day when about the only things Canada exported were natural resources and tractors, half of Clairtone's stereos were being sold in the U.S., at "prestige accounts" like Abraham & Straus and Bloomingdale's in New York, Marshall Field's in Chicago, Halle Brothers in Cleveland, and J.L. Hudson's in Detroit. For a Canadian consumer product to be featured in the windows of Bloomingdale's was almost unimaginable—and yet, there it was.

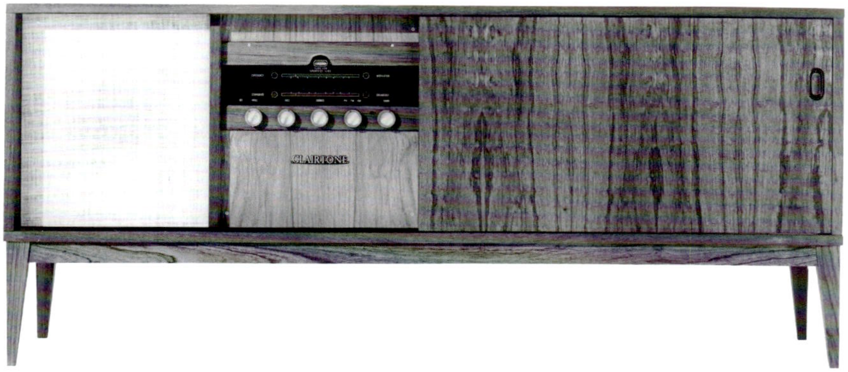

Clairtone's first prototype, the monophonic 100-M of 1958 with a 20-watt, 15-tube amplifier from Granco, and a Dual turntable, never went into production. It was replaced by the stereophonic 100-S.

Peter Munk Associates, Clairtone's predecessor, made custom hi-fi sets for wealthy Torontonians. The cabinetmaker George Weil, kneeling, would become one of Clairtone's first employees.

In 1959, when fashionable men, including my father and David Gilmour, still wore three-piece suits, an article in the *Globe and Mail* trumpeted Clairtone's success in the U.S. market: "Canadians would have popped a few buttons on their vests last week if they had attended the American Music Show in New York. A stereophonic set designed and manufactured by a Canadian company founded less than a year ago by two young Canadians was the centre of attraction.... This is perhaps the first time a piece of Canadian consumer-electronic equipment has aroused such enthusiasm in the U.S."

As for my father and David, they were hailed as visionaries. They were "everybody's darlings," in the words of the journalist Alexander "Sandy" Ross. "They were treated as movie magazines treated Rock Hudson, with awe-struck approval," another journalist recalled. "Peter Munk was probably one of the most admired young men in Canada, the closest thing to a hero the Canadian business community has produced in this generation," continued Ross. "Just contemplating the Clairtone phenomenon made us all feel smart and groovy and efficient, like the Scandinavians almost."

Even my father seemed awe-struck by his own success. "There was a year when I had 34 speaking engagements," he recalled wide-eyed on CBC's *the fifth estate* in 1978. "I stood there, at the age of 30, lecturing the stalwart, establishment members of the Canadian business community. I used to go home ... and pinch myself."

MY father has an innate sense of timing, almost a sixth sense. In 1965, when a reporter asked him to explain Clairtone's success, he remarked: "The idea was good. But the idea would have been completely lousy had it happened five years earlier or five years later."

The cover of one of Clairtone's first brochures, designed by David Gilmour and starring his wife, Anna, on Toronto's Cherry Beach, 1958.

For one thing, in the late 1950s, the way people listened to music was changing. Transistors, stereophonic sound, long-playing records—all were being introduced or becoming popular just as Clairtone got into business. At the same time, the style we now call "Mid-Century Modern" became fashionable and a sign of educated good taste. Here and there, slowly, Canadians began replacing their bulky chesterfields and Victorian credenzas with furniture by Charles and Ray Eames, Eero Saarinen, Finn Juhl, and Hans Wegner, to name only a few of the international designers who left their mark on Canada.

By the late 1950s, in Toronto, a handful of shops were dedicated to modern Scandinavian design. Georg Jensen, on Bloor Street, was across the road from Shelagh's of Canada, whose owner, Shelagh Vansittart, was David Gilmour's older sister. Further south, on Bay Street, Avrom Isaacs opened his first art gallery in 1956. Before you knew it, a group of young Canadian abstract expressionists—Graham Coughtry, Michael Snow, Joyce Wieland, Gordon Rayner, Richard Gorman, William Ronald, and John Meredith—were stars in the Toronto sky. Meanwhile, the Finnish architect Viljo Revell won an international competition to design the new Toronto City Hall. Ludwig Mies van der Rohe, having just completed his Seagram Building in New York, began working on another modern icon, the Toronto-Dominion Centre.

Don't misunderstand me: Clairtone wasn't on the cutting-edge in the late '50s. My father and David thought of themselves as industrialists, not artists. That said, Clairtone's success rode in on the wave of modernism. Clairtone's 1960 annual report, printed in the era's colours of mustard yellow and avocado green, suggests how completely my father and David had embraced contemporary design. They trusted it.

Not many publicly traded companies in 1960 would have thought to hire Chris Yaneff to design their annual report; and yet Yaneff was only one of many innovative designers, graphic artists, and typographers who worked for Clairtone. Carl Dair, Hugh Spencer, Frank Davies, Anthony "Tony" Mann, Al Faux, and Burton Kramer: all of them played a role in shaping Clairtone's image—in crafting the company's products, logos, brochures, advertisements, and hang-tags, thereby creating a unified, "holistic" body of uncluttered work that still looks modern and relevant.

One of Chris Yaneff's best-known designs is the classic, minimalist, Windsor Salt box, a composition of blue and brown circles that for some

Munk with Gilmour, seated in the Egg chair. This look became the signature image of Clairtone's early marketing campaign, created by Dalton Camp and Chris Yaneff.

Profitable PR / *Chris Yaneff*

IT WAS TIME

Chris Yaneff tells the story behind the PR campaign that helped Clairtone rise from nowhere to become a household word

When Business Week in 1962 was preparing to run a major story on the Clairtone Sound Corp., the editors asked how an unknown new company — with a shoestring budget, operating in a fiercely competitive market against such already established giants as CGE, RCA, Philips, Westinghouse and a half dozen others — managed to accomplish so much with so little.

The answer was simple: Take the electronic skill of engineer Peter Munk, add the creative ability and flair of designer David Gilmour, and you have the basics for one of the greatest Canadian sales stories of a decade.

Almost out of the picture, but there just in case something goes wrong, PR-man Chris Yaneff cites the official attention given to the new Clairtone plant opening as a prize example of a successful PR function. Group included, left to right: Peter Munk, Clairtone president; former Toronto mayor Nathan Phillips; R. W. Macauley, then Ontario Economics and Development minister; and George Hees, then federal minister of Trade and Commerce.

Attending Clairtone's factory opening in Rexdale, Ontario, in 1962, from left to right: Peter Munk, Nathan Phillips (mayor of Toronto), Robert Macauley (Ontario minister of Economic Development), and George Hees (federal minister of Trade and Commerce). Over on the far left is designer and "PR man" Chris Yaneff, who organized the event.

people recalls a painting by Piet Mondrian or Ellsworth Kelly. Yaneff's 1960 annual report for Clairtone is equally significant. On the cover is one small, isolated black-and-white image—a photograph of my father and David, both in dark suits with white pocket handkerchiefs, their black shoes polished to a high gloss.

No Clairtone product appears on the cover of that annual report. Instead, David is seated casually, confidently in Arne Jacobsen's classic Egg chair of 1958—black leather with a star-shaped aluminum base—engaged in conversation with my father. The Egg chair is there to define and identify the company, to suggest by association and a leap of imagination that a Clairtone stereo, like Jacobsen's Egg chair, is forward-thinking and timeless.

DAVID and my father also posed for a series of full-page advertisements, which promoted them as the "internationally-acclaimed creators of Clairtone stereophonic high fidelity units." The ads ran in all the major Canadian and U.S. newspapers, as well as in the *New Yorker*, *Time*, and in playbills for Toronto's Royal Alexandra Theatre. Later, when Clairtone began making colour TV sets, David and my father starred in television commercials.

"Peter Munk and David Gilmour are *very* successful Canadians," intones the voiceover in one Clairtone TV commercial, directed by the Canadian cinematographer Fritz Spiess. Seated in a restored 1936 Pierce-Arrow convertible, my father and David are shown driving leisurely across the Brooklyn Bridge into Manhattan, on their way to B. Altman's, the grand department store on lower Fifth Avenue, to deliver a Clairtone television

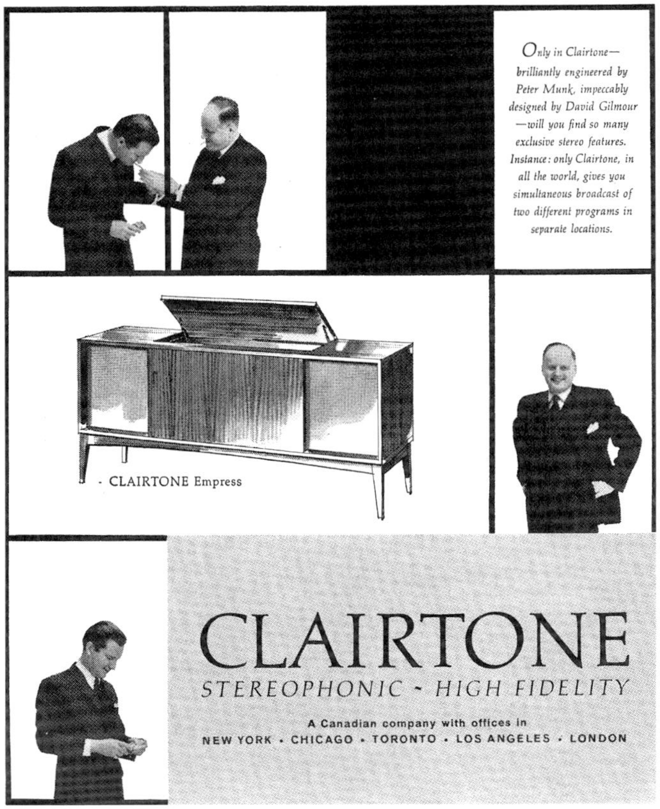

An early example of Dalton Camp's award-winning campaign for Clairtone, showing the 600-S (the Empress), introduced in 1960. Camp played up the youth and bravado of Munk and Gilmour.

set. According to the 1967 production notes: "As men carry set through to entrance of store, as per board, Gilmore [sic] moves into frame and dusts top of set with hanky to make the whole action appear extremely important."

Today, in the era of YouTube, MySpace, Flickr, and Facebook, the art of self-promotion is hardly worth talking about. But 50 years ago, when understatement was still considered a virtue in Canada, my father's and David's showmanship was electrifying, and shameless.

"The boys," as they were often referred to back then, were a perfect team. David, pedigreed, tall, and lean, brought a refined taste and a touch of class to the operation. In Clairtone's ads, his signet ring is clearly visible, like a stage prop. He's identified as "David Gilmour, Designer," though in truth he was not so much a designer as a stylist.

By contrast, my father ("Peter Munk, Professional Engineer"), who was not especially lean in those days, is portrayed as the mastermind—the intense, brooding genius, the size of his head exaggerated by a receding hairline. He's often shown with a cigarette. If David was meant to appeal to female shoppers (and he was), my father played the sober engineer, deeply concerned with millivolts and megahertz.

Many of Clairtone's print advertisements were written by Dalton Camp, the masterful Canadian political strategist who died in 2002. Not surprisingly, Camp handled Clairtone's award-winning campaigns as though David and my father were running for office. "Messrs. Munk and Gilmour are so effective a team they have scored a remarkable international triumph," reads one of Camp's typical advertisements, its inflated prose suggesting that something more, something grander than a stereo is being sold. "Why has there been such acclaim? We invite you to see and hear Clairtone yourself. Note the purity of line, unmarred by gild or chrome embellishment, and—even more important—note the brilliant clarity of reproduction."

Interviewed for Garth Hopkins's 1978 book, *Clairtone: The Rise and Fall of a Business Empire*, Camp recalled: "Over the course of two or three meetings with Peter and David, I became so totally imbued with and dedicated to what they were trying to do, that I decided to handle the campaign myself....

"I sold them on the personalized and highly individualized marketing concept for Clairtone. That was relatively easy to see as the right thing to do, because both of them had always tended to sell themselves first and their concept or product second."

TC: Mr. Hugh Spencer

FROM: Peter Munk

November 4, 1963.

Re: Company Stationery.

Thank you for your recent memorandum concerning requests for individualized company stationery.

The answer to any such requests must be in the negative. My instructions on this subject are that no one but the President of Clairtone Electronic Corporation and the President of Clairtone Sound Corporation Limited is to use anything but the regular Clairtone stationery. I think everyone should be proud of the distinctive design of this letterhead and I feel strongly that personal identification only detracts from the Clairtone name.

PM/C
c.c. Messrs. Hahn, Pols, Berenyi, Cohen, Freeman.

MEMO TO: Mr. David Pols — cc: Mr. P. Munk ✓, Mr. Z. Hahn

FROM: Hugh Spencer

DATE: December 31, 1963

SUBJECT: GENERAL ADVERTISING AND PROMOTION BUDGET 1964

This week we enter fiscal 1964 and, in absence of information to the contrary, I assume that I now am responsible for the administration and application of the 1964 budget as above.

No discussion of the figures presented by myself for your consideration has been held. I would appreciate your formal acceptance of this budget and your clearly defined authority to purchase, under my own supervision, within the limits of my budget presentation for this Department.

HS/jb

MAR 3 1964

MEMORANDUM TO THE STAFF

FROM: Hugh Spencer

DATE: March 3, 1964

SUBJECT: CONFIDENTIAL WORK

The work carried out in the Advertising and Design Department is confidential. Considerable interest in the activities of this Company is held by our competitors.

The offices of the Design Department and the desks used by Anna Spencer should be regarded as areas that are not common ground.

You are particularly requested not to bring visitors to the Design Department without prior appointments and on no account may any member of the Company spend time in this Department without proper business.

HS/jb

CLAIRTONE — INTER-OFFICE MEMORANDUM

CLAIRTONE SOUND CORPORATION LIMITED
Toronto New York Chicago Los Angeles London

May 29, 1964

TO: Hugh Spencer

FROM: Peter Munk

RE: Re your memo of May 15th regarding Department Head Meetings.

It is this kind of memo that is one of the causes of tremendous inefficiencies in this Company, i.e. use of time, paper, typists' and executive time in reading them.

It would be appreciated if you would kindly confine your sentiments to verbal discussions with Management and Staff.

PM/JB

Hugh Spencer staked his claim to Clairtone's design department in a series of inter-office memos, some of which are shown here. His territorial ambitions were thwarted by Peter Munk.

David Gilmour (right) at the Los Angeles Design Center, with the building's architect, Richard Dorman.

BRUCE Mau, the respected Canadian designer and an admirer of Clairtone, suggests that my father and David viewed good design as a marketing gimmick, not a philosophy. He may have a point. After all, Clairtone's most-celebrated product, the Project G stereo, was originally conceived as an elaborate promotion, a kind of concept car that never goes into production. Introduced at the National Furniture Show in Chicago in January 1964, Project G was intended to give Clairtone an aura of futuristic cool. In other words, it was meant to promote the company's other, more prosaic hi-fi sets.

Initially, Clairtone planned to make only a handful of Project Gs. At some point, according to Clairtone documents catalogued at Design Exchange and at the Nova Scotia Archives, that number was increased to 150, then 400. Exactly how many Project Gs were actually made during the short time it was in production is not clear, but based on the information we have—and given the fact that Project G was impractical, expensive, and hard to produce—we can assume that no more than a few hundred were ever manufactured.

The whole idea behind Project G was to create a product with "cachet," desirable but beyond the reach of most people. In the first place, it was priced at $1,850, equal to about $12,500 today; that's comparable to the cost of a small car in the 1960s, and it's more than twice the price of what was then Clairtone's most-expensive stereo. What's more, the G didn't look like any other hi-fi set. It was striking, and it was massive: nearly seven feet long (over two metres), it was too big for an ordinary living room. Then again, it wasn't designed for ordinary persons.

"Trend setters" were the target market for Project G, according to an internal company document from 1963. "It is these people who make new painters famous, who adopt new styles in furniture, who drive exotic cars, spend their holidays in never-before-heard-of places, wear new styles of clothing and acclaim musical creations before any one else has heard of them," the document explains, knowingly. "It is a small and very elite group."

To make Project G an object of desire, Clairtone hired Irving Penn to photograph Project G the way he'd photograph a seven-foot-tall fashion model. Then the company spent even more money on 25,000 promotional booklets, distributed to all the best-known designers, architects, decorators, and "qualified customers" in the U.S. and Canada.

A publicity shot photographed at 63 Woodlawn Avenue West in Toronto, a house designed in 1961 by the architect Jerome Markson for the author's parents, Peter and Linda Munk. Note Franco Albini's Margherita chair.

Modern and defiant, Project G was intended for "trend setters."

In New York, Clairtone hired William Safire's well-known public relations firm to choreograph the launch of Project G in the United States. In Los Angeles, an agency specializing in product placement arranged for Project G to appear in *Marriage on the Rocks*, the 1965 comedy starring Frank Sinatra and Deborah Kerr, and in *I'll Take Sweden*, with Frankie Avalon and Tuesday Weld. One after another, like bait, photographs of Hollywood celebrities posing with Project G were released to the media.

So carefully managed was the introduction of Project G that David Gilmour himself composed an eight-page guide advising Clairtone's sales reps on the finer points of eliciting free publicity. Point number C.2(f) warned them: "You may have to stage a 'publicity stunt' (a term you shouldn't use outside the Clairtone family)."

HUGH Spencer started designing Project G in 1962, he recalled in an interview with *Marketing* magazine. As he explained, my father had simply asked him to "design something 'way out' for next year."

Spencer's first maquette was a small wooden box with two tennis balls attached, one on either end. Even on that scale, the design was radical. The speakers weren't only outside the cabinet, they were *round*. Who had ever seen round speakers?

Recently, I asked my father and David if they remembered the first time they saw Spencer's model of the G. "I remember the day," said David. "We saw these two round tennis balls..."

"Mind-blowing," interrupted my father.

"We said 'This is it,'" continued David. "We fell in love with it the instant we saw it."

vol. 11, no. 6 — june, 1964

PLAYBOY

Copenhagen P. 81
Jackpot P. 70
Mamie P. 110
Gifts P. 118

CONTENTS FOR THE MEN'S ENTERTAINMENT MAGAZINE

HUGH M. HEFNER *editor and publisher*

A. C. SPECTORSKY *associate publisher and editorial director*

ARTHUR PAUL *art director*

In a major coup for Clairtone, Project G was included in the June 1964 issue of Playboy. Stan Freeman, a Clairtone sales executive, delivered a Project G to Hugh Hefner at the Playboy Mansion: "Hefner loved it, and agreed to feature it in his magazine and play it during his parties."

"We had no doubt we'd discovered the new shape of sound," said my father.

By the time Project G actually went into production, dozens of improvements had been made to Spencer's original design. For example, because David thought the original cabinet looked like a coffin, he wanted the sides to be tapered. And while Spencer wanted to use teak for the cabinet, George Weil, Clairtone's head cabinetmaker, pushed for rosewood instead ("I thought teak looked too plain," explained Weil). Other changes were made: for example, a clear plastic lid was replaced by a wooden tambour roll-top; real leather side panels were replaced by leatherette; and the scale of the cabinet was altered to accommodate the stereo components.

The end result, the final Project G, is now a design icon: modern and cool and defiant. In some respects, it suggests Charles and Ray Eames's luxurious rosewood-and-black-leather lounge chair of 1956, famously described by Charles Eames as having "the warm receptive look of a well-worn first baseman's mitt." With its aggressive black aluminum spheres, Project G may be more outrageous than the Eames chair; certainly it's less receptive.

In all events, from the moment it was released to the public in 1964, Project G was praised by critics, by the media, and by "trend setters." That year, 1964, Project G won a silver medal for excellence in design at the Milan Triennale. Oscar Peterson affirmed publicly that his music sounded as good on a Project G as it did live. And Hugh Hefner, no stranger to objects of desire, acquired a Project G for the Playboy Mansion.

Project G epitomized the Swinging Sixties, as defined by Hefner. *Playboy*'s June 1964 issue, featuring "Gifts for Dads and Grads," showed Project G alongside other masculine status symbols: Børge Mogensen's oak-and-leather Spanish chair; a Mark Cross cigarette lighter with "trigger-action release"; the new Ford Mustang convertible (with white-wall tires); and High Standard's Supermatic Trophy .22 target pistol.

Even after Clairtone's P.R. department had been disbanded, Project G and its successor, the G2, continued to appear in movies and TV shows. In 1967, for example, the G2 appeared in *The Graduate*, with Dustin Hoffman and Anne Bancroft. Most recently, in 2006, it showed up in *Invincible*, an otherwise tedious film starring Mark Wahlberg. Among collectors, the G series has taken on a cult status.

Project G's designer, Hugh Spencer, moved from England to Canada in

A G2 with Greg Kinnear and Paige Turco in *Invincible* (Walt Disney Pictures, 2006).

1956 and worked with Clairtone from the beginning. In the late 1950s, his name appears on an early accounts payable list for Peter Munk Associates, my father's custom hi-fi company, Clairtone's predecessor. Spencer's first big job for Clairtone was creating its elegant arrow-like logo in 1961—a simple, asymmetrical white arrow set on a black background, and later updated by the graphic designers Frank Davies and Burton Kramer.

Spencer was a great talent; everyone I've spoken to acknowledges that much. At the same time, he was hard to work with: touchy, hotheaded, and often grandiose. "It took the patience of Job to work with him," is all David Gilmour would tell me about Spencer (on the record, that is). George Weil, Clairtone's head cabinetmaker, sighed when I asked about his work with Spencer: "Oh, we had some fights, I can tell you that." According to my father: "Hugh was temperamental. He would sulk."

For all that, my father never lost sight of Spencer's genius as a product designer. "I needed Hugh Spencer," he explained to me. "I couldn't afford to lose him."

WHAT very few people know is that by 1964, the year Project G was launched, Clairtone was desperate for its next big act. For all the razzle-dazzle—the award-winning advertising campaigns, the celebrity endorsements, the media coverage—Clairtone was losing momentum. Sales, while still impressive, weren't growing as quickly as my father and David had anticipated. Profit margins were falling. Internally, people grumbled about the huge sums being spent on advertising and marketing; more to the point, they wondered if all the advertising and marketing campaigns were really effective. Meanwhile, as attested by a stack of angry

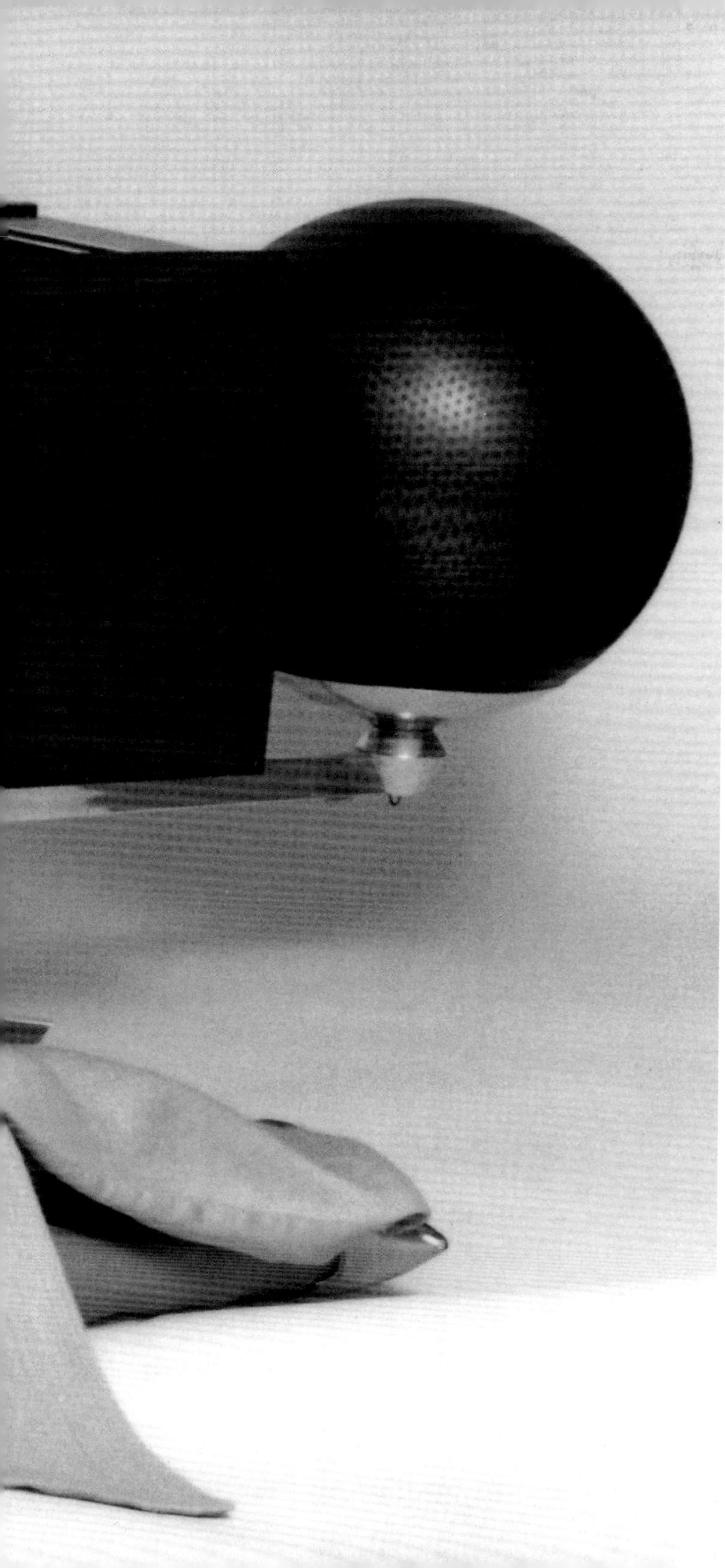

Tuesday Weld and Project G, 1965.

letters held in the Nova Scotia Archives, Clairtone was threatening to sue its largest supplier, General Instrument of Canada.

Some of the company's biggest problems were south of the border. Portrayed publicly as the engine of Clairtone's glorious future, business in the U.S. was actually veering off track. In a four-page, single-spaced letter, dated October 18, 1963, my father scolds David (who was then running Clairtone's U.S. subsidiary) for the sudden sharp drop in U.S. sales. Separately, in a confidential memo to my father, Clairtone's chief financial officer, Zigmund Hahn, advises that "emergency measures" are needed: "I do not need to remind you that the U.S. market organization from the beginning was distinguished by a series of indecisions, compromised policies and inadequate personnel." No wonder the launch of Project G, especially in the U.S., took on such urgency inside the company. No wonder Hugh Spencer became invaluable.

Something else: Clairtone was desperately, recklessly short of cash. It's hard to convey just how difficult it was to raise capital back then. There were no angel investors in Canada. No incubator funds or venture capitalists were knocking at Clairtone's door, and no one used credit cards to fund start-ups.

In 1957, my father got into business the old-fashioned way: his father-in-law, my grandfather William Jay Gutterson, gave him $3,000 of capital. Later, in the early days of Clairtone, David Gilmour kept the company going by taking out a mortgage on his house. When more capital was needed in 1960, Irving Gould, a Toronto financier, raised $250,000 for Clairtone by selling 12 per cent of the company on the unlisted, or over-the-counter market.

But the bigger Clairtone became, the more money it needed, until, finally, my father's chief preoccupation was finding ways to keep the company solvent, to save it from one financial crisis after another. "It was an impossible situation," my father remembered. "I was down at the banks every day. But the banks were tough as hell with young companies. I couldn't even get short-term money. It was impossible. Yet if we'd stopped growing, we wouldn't have been able to stay ahead; we'd have killed ourselves. It was an impossible, an absolutely impossible situation."

By 1964, in what would seem to be counterintuitive thinking, my father concluded that the only way to break the double-bind was to get bigger faster. Reasoning that the RCA Victors and Magnavoxes of the world did not struggle with day-by-day cash flow problems, he decided to make Clairtone

A chartered plane brought important customers to Stellarton, Nova Scotia, for the factory opening.

The assembly line in Stellarton.

an industrial giant. And that in turn meant building a huge modern factory to bring every part of the manufacturing process, from the cabinets to the electronic components, in-house and under one roof. Of course there was one great problem: If Clairtone could barely fund its day-by-day operations, what were its chances of financing an $8-million factory?

THERE'S something you should know about my father: the more impossible the situation, the more single-minded he becomes. "He's always crossing the Rhine," as my mother puts it, by which she means that my father operates best on the offensive, in an assault boat, as it were, fighting a hostile takeover, challenging a board of directors, and plotting his next move.

George Weil, who worked for my father between 1957 and 1968, remembers flying with him in a small, single-engine prop plane to visit Clairtone's cabinet plant in Strathroy, Ontario. "I'll never forget that flight," Weil told me. "We flew into a severe thunderstorm and the pilot said, 'We have to turn around.' But Peter Munk said, 'What kind of pilot are you? Keep going!' He was insistent. I'll never forget it." The pilot continued on course.

In 1964, when his dreams of building Clairtone into an industrial giant seemed far-fetched, even ludicrous, my father convinced the government of Nova Scotia to build him his factory. For both parties it was a matter of self-preservation: Clairtone would get $8 million of (almost) free money in exchange for the small price of moving its operations to the depressed backwater of Pictou County, Nova Scotia. "This arrangement is absolutely fantastic!" my father declared at a press conference announcing the deal. "I can't understand why more companies don't take advantage of it."

On June 21, 1966, in the presence of Robert Stanfield, the Premier of

Peter Munk delivers his famous (or infamous) speech at the opening of Clairtone's factory in Nova Scotia, June 21, 1966: "We shall be at the forefront of your industrial revolution!"

Nova Scotia, John Turner, the Federal Minister of Finance, and nearly 1,000 distinguished guests, including my parents and David Gilmour, the Clairtone factory opened officially. Stretched over seven acres, it was a marvel of high technology.

Every last detail of the factory was designed to improve quality and productivity. An automated dust-collection system took in air at the rate of 90,000-cubic-feet per minute. A miraculous seven-headed moulder, made in Germany, performed seven woodworking operations at once, including joining, moulding, sizing, fluting, and grooving. "It can take a piece of raw wood and produce a basically completed French provincial leg," marvelled one trade publication. Even the factory lighting contributed to worker efficiency; according to research cited by Clairtone, the new, intense bulbs had been "shown to increase productivity by eight percent, greatly reducing fatigue and cutting down errors."

At the opening ceremony, pipers piped. Residents of Stellarton, the town where the new factory was situated, cheered: they'd been promised at least 600 jobs to begin with, and up to 2,000 jobs in the bright future. In a mesmerizing speech that was broadcast right across Canada by the CBC, my father offered a heartfelt pledge to Premier Stanfield: "Sir, we will not let you down! We shall be at the forefront of your industrial revolution!"

ASK my father and David about the collapse of Clairtone, and you'll be told that the move to Nova Scotia killed their company. That's true as far as it goes, but it may be more accurate to say that Clairtone was undone by their tendency to overreach. "Munk was too good a salesman for his own good," to quote William Mingo, chief counsel for Nova Scotia's Industrial Estates Limited (IEL), the Crown corporation that funded Clairtone. "He could sell anything to anyone—including himself.... My, he was a promoter. My, he had energy. My, he had charm. My, he had imagination."

One thing is certain: Clairtone built one of the biggest, most modern factories in the Western hemisphere in a place entirely unsuited to manufacturing. "Stellarton's prospects were bleak even by Nova Scotian standards," writes Harry Bruce in his 1985 biography of the businessman Frank Sobey, who served as mayor of Stellarton and president of IEL. At one point, half of Stellarton's population had been "wholly dependent" on the coal

Clairtone's G-TV, introduced in 1966, in a typical interior of the time.

industry. Then the mines shut down. "Without mining, the town was like a movie house without a projector," writes Bruce.

By the time my father and David acknowledged the obvious, it was too late. "The general population is basically not geared to the manufacturing frenzy and especially the five-day workweek," reported an intensive study of the factory, quietly commissioned by Clairtone in 1967. "The welfare situation is such that it has created conditions similar to Appalachia in the United States where the third generation is already on relief."

The labour force in Stellarton was only one of Clairtone's problems. Nothing went smoothly, from all accounts. Cost controls were "nonexistent," according to the commissioned study. Roads leading to the factory were so bad that Clairtone's stereos and TVs were often damaged en route. Getting parts for the stereos, a task that at the old plant in Rexdale, Ontario, could be organized in a single day, took several weeks. As for the manufacturing process itself, Clairtone's inexperienced management team was over its head, and sinking: inventory went missing; unidentified parts accumulated on the factory floor; the assembly line seldom if ever ran at full speed or full capacity.

In the Clairtone archives, one memo after another foreshadows the catastrophe. "It is unbelievable and very disturbing to find that at this late date, when we are scheduled to complete 1,500 of the T-14 chassis, the flywheel assemblies have been delivered for only 550 chassis," reads one urgent note from Michael Chojnacki, Clairtone's vice president for manufacturing. "We are still unable to commence delivery of TV sets," reads another memo. "Instead, the very few G-TV's once again have been condemned due to electrical and mechanical problems."

By the summer of 1967, even my father was in a panic. "Chaos" is how he would later describe the situation, in his testimony before the Nova Scotia legislature in 1969. Instead of showing a profit for the year, there would be a huge and staggering loss. Anxious suppliers, figuring they'd never get paid by Clairtone, stopped shipping parts. The once-promising business of colour TV was a bust. And yet again, Clairtone was in dire financial straits.

Only a few months earlier, in late 1966, IEL had invested another $3 million in Clairtone. This time round, however, not even the Nova Scotia government was willing to throw good money after bad. For months, relentlessly, my father tapped every connection he had to raise money for Clairtone. Through the brokerage firm W.E. Hutton & Co., he sold $2 million

A G-TV with Peter Sellers and Claudine Longet in The Party (MGM, 1968).

Detail from a Dalton Camp advertisement.

worth of convertible debentures; to survive he needed more than that. He tried getting Clairtone listed on the American Stock Exchange. He also came close to selling Clairtone to the Singer Company, the giant U.S. conglomerate. Still, for all his ingenuity … nothing.

There's something I haven't told you. Just at the moment when Clairtone's hi-fi and TV business most needed his attention, my father was chasing the next big thing: the automotive industry. In 1964, when very few North Americans had ever seen a Japanese car, he and a brilliant Australian, Pat Samuel, had bought control of Canadian Motor Industries (CMI), whose one asset was the exclusive right to assemble and sell Toyota and Isuzu cars in Canada. Depending on your point of view, that deal—a complicated manoeuvre that resulted in Clairtone's, and the Nova Scotia government's, brief ownership of CMI—is: (a) an example of my father's far-sightedness, or (b) an example of his recklessness.

One way or another, what matters in a book about Clairtone is that my father's short-lived ambition to become an automotive magnate hastened Clairtone's downfall. There simply wasn't enough money or managerial talent to go around. "I've never worked as hard as I did that year," my father told me recently, referring to the period between mid-1966 and mid-1967 when Clairtone, along with CMI, fell apart. He's now 80 years old and still remembers every humiliating detail of the final months at Clairtone. "Everything I've done since then has been easy."

On August 27, 1967, 11 years, two months, and 18 days after founding Clairtone Sound Corporation, my father and David Gilmour lost control of the company to the government of Nova Scotia. "A Canadian Success Story Turns Sour," was the headline in the *Toronto Daily Star*. It was over.

By the early 1970s, to lure shoppers into their grocery stores, Loblaws

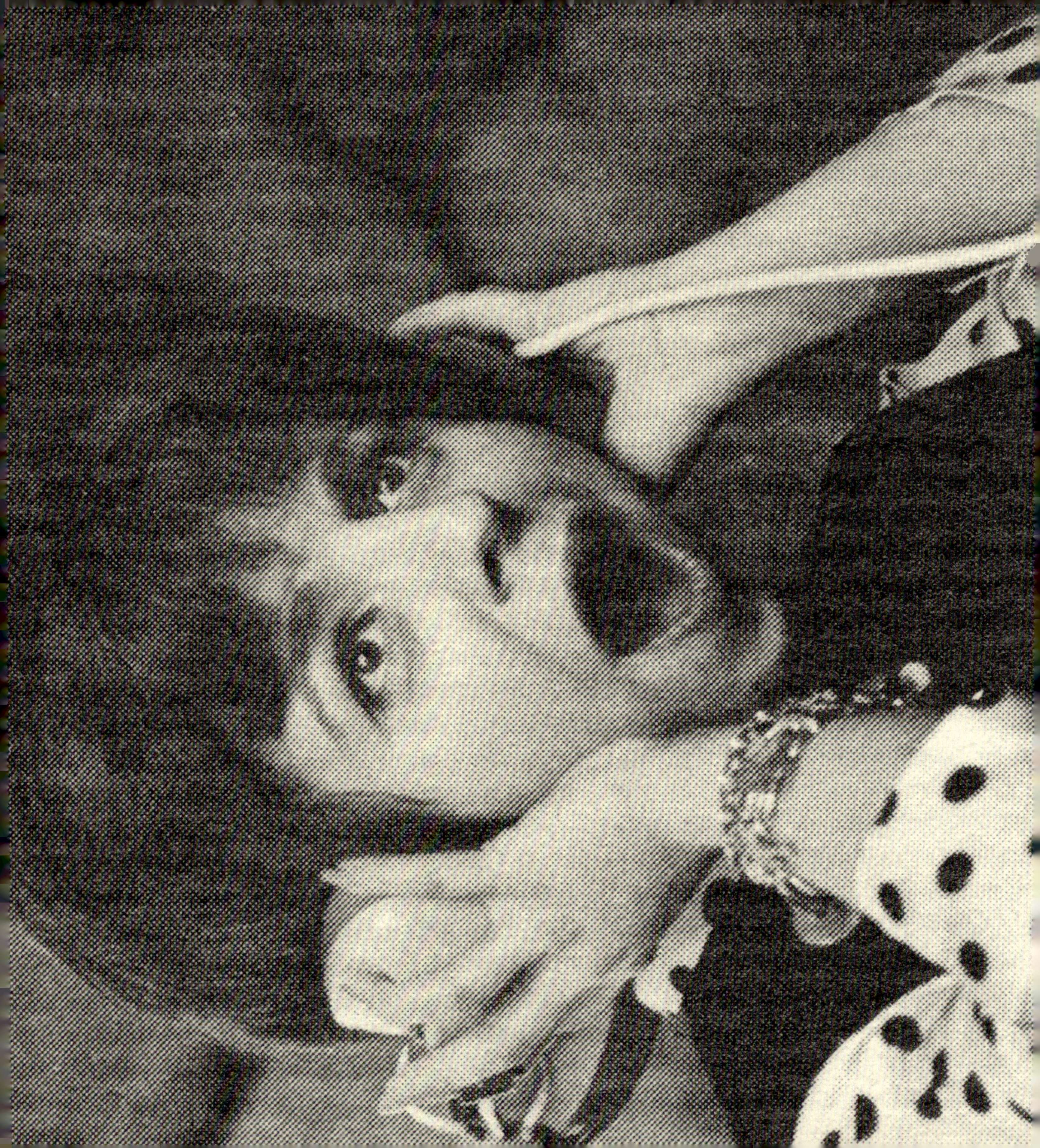

At Clairtone's showroom in Rexdale, the jazz singer Carol Sloane listens to a Project G with "stereophonic headphones."

was running newspaper ads for games of "Double Money Bingo"; the grand prize was a free Clairtone colour TV. That's how far the company had fallen since the heady days of Frank Sinatra, and Irving Penn, and Pierce-Arrow convertibles in New York City. In 1971, Clairtone was de-listed from the Toronto Stock Exchange. By the time the company officially closed its doors in 1972, it had become the antithesis of everything my father and David had imagined.

Nina Munk and her father, 1967.

I WAS born in September 1967, less than two weeks after my father lost Clairtone. My parents sold their house a few months later, and we moved into a rental apartment in Toronto. "I ended up with less money than the mortgage was worth on my house," my father later told *Executive* magazine. "After being sort of a hero in Toronto for ten years and considering yourself to be a millionaire, suddenly your friends are thinning out and the same banks don't talk to you in the same way and you don't get invited to the same boardrooms and no universities are asking you to give the keynote address for their business graduate course. Then comes a three- or four-month period when you really wonder whether it was all just a fluke."

It was not a fluke, of course. My father became a serial entrepreneur—one of Canada's most celebrated businessmen. When Clairtone collapsed, however, not many people were betting on him. He'd lost his credibility, and they'd lost faith. An exception was Frank Sobey, the empire builder from Nova Scotia who, as president of Industrial Estates Limited, made the grand mistake of backing my father in the first place.

In 1968, when an inquiry into the Clairtone affair was held in the Nova Scotia legislature, Sobey was called to testify. Even as he came under attack, he defended my father. "It's a good thing we have people like Peter Munk in Canada. He's a builder. He has the ability, he has the energy, he has the courage to go out and create industries. It's people like Peter Munk that created all our industries in the United States and Canada.... Invariably they lost most of their capital in doing it, because they went too fast and too swift."

Sobey wasn't finished. "It's people like that that trigger industrial development, people like Peter Munk that have courage. The people that sit back without courage, and do a lot of talking and a lot of criticism, make no contribution to the industrial development of our country. That's all I want to say."

LEFT Clairtone was a little-known brand in January 1960 when this Simpson's advertisement appeared in Canadian Homes and Gardens. The department store was Clairtone's biggest retailer.

TOP RIGHT This staged photo of Clairtone's production line appeared on the cover of the Toronto Stock Exchange Review in 1963.

BOTTOM RIGHT A 1964 advertisement shows the range of products, from old-fashioned to avant-garde.

ERA ONE

Founded in 1958, Clairtone grew with dazzling speed. Within its first three years of operation, it opened a factory in Rexdale, Ontario, bought two furniture plants in Strathroy, Ontario, and established itself in the U.S. with showrooms in the Time-Life Building in New York City, the Merchandise Mart in Chicago, and the Design Center in Los Angeles.

Right from the start, Clairtone showed a strong commitment to modern design, introducing aesthetic and technological innovations to its product line: Scandinavian-styled, low-slung cabinetry made of oiled teak and walnut; clean, graphic faceplates; sculptural dials and knobs; earphones for "silent listening"; wireless remote control; and tubeless, "solid state" circuitry. The company employed some of the most talented designers and original thinkers working in Canada at the time—among them, Carl Dair, Dalton Camp, Chris Yaneff, and Hugh Spencer—who ensured that Clairtone used the wittiest copy, the latest typefaces, and the most up-to-date exhibit and signage systems. In 1961, Clairtone boldly partnered with Braun, the German electronics firm, to sell a line of modern radios and phonographs under the Clairtone-Braun name.

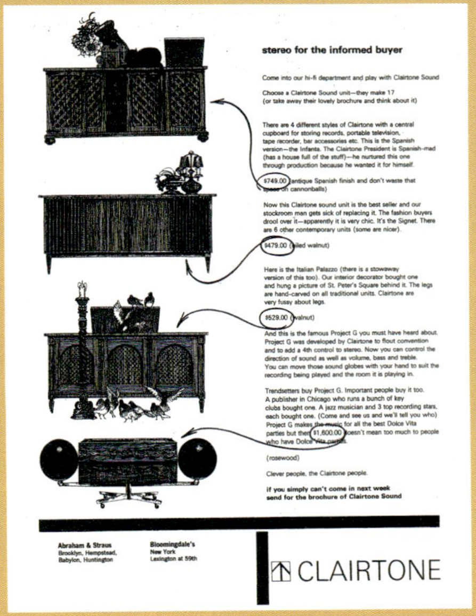

For all that, Clairtone's executives soon came under pressure from dealers who wanted a more conservative, more mainstream product line. "We went into Italian provincial and French provincial—things that made me sick to have to do," David Gilmour recalled. "But as we grew we couldn't just thumb our nose at the market." And so, Clairtone produced more and more stereo consoles with regal names like "the Baroness" and "the Duchess." The design of these cabinets clashed with the tenets of Modernism, but faux styling sold, and sold well.

With Spencer's help, Munk and Gilmour found a way to keep Clairtone avant-garde. The Project G stereo, with its twin round speakers (or "sound globes," as the company's P.R. department preferred to call them), became Clairtone's most-celebrated design. There was nothing like it. Project G's radical new look anticipated one of the popular motifs of '60s space-age design—the sphere, used in visionary architecture and interior design as well as in consumer products.

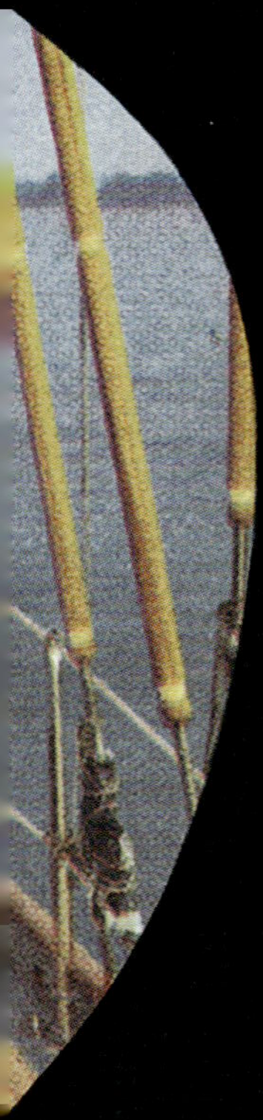

CLAIRTONE

THE IMAGE OF GRACIOUS LIVING
Even in the early days, with almost no money for marketing, Peter Munk and David Gilmour were determined to position Clairtone as a luxury brand. To attract a high-end clientele, it was important to market the image of "gracious living" as promoted by 1950s etiquette books—in this instance, a beautiful girl sailing on Lake Ontario with Clairtone's 200-S stereo console. David Gilmour art-directed Clairtone's early brochures, including this one from 1958. He borrowed the boat from his friend Harry Kohl, an architect, and convinced his wife, Anna, to be the model. "I wore a dress that I saw in *Vogue*; David ordered it for me," she remembered. "I used to wear it when we went dancing at the Royal York Hotel." As a final touch, Gilmour used a piano-shaped stencil that he found at a local art-supply store.

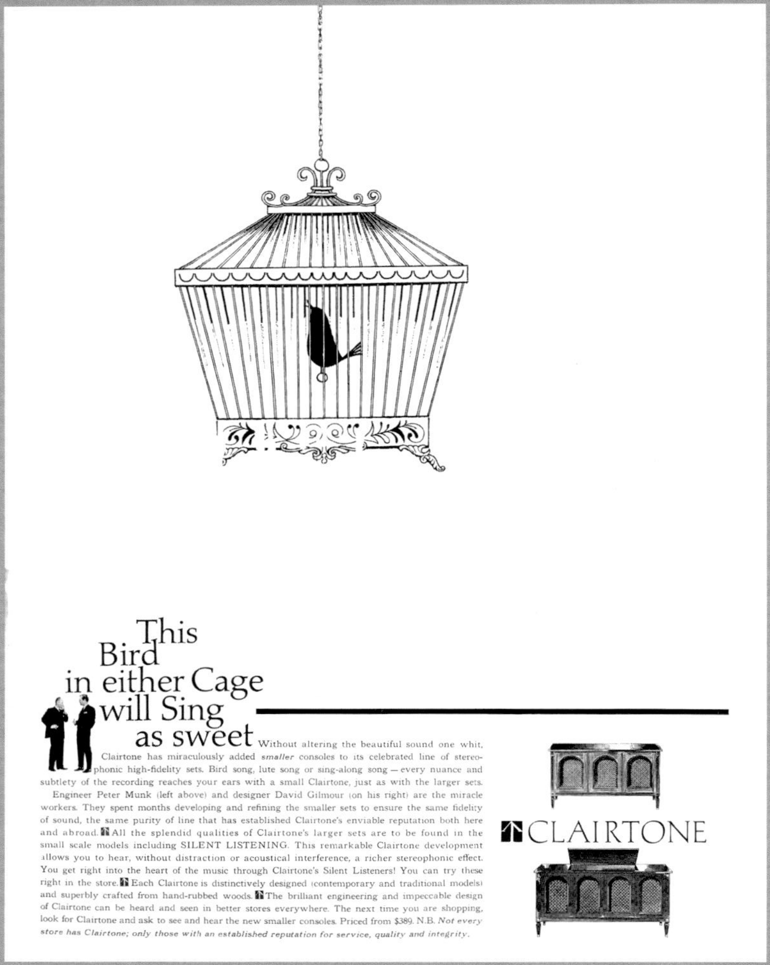

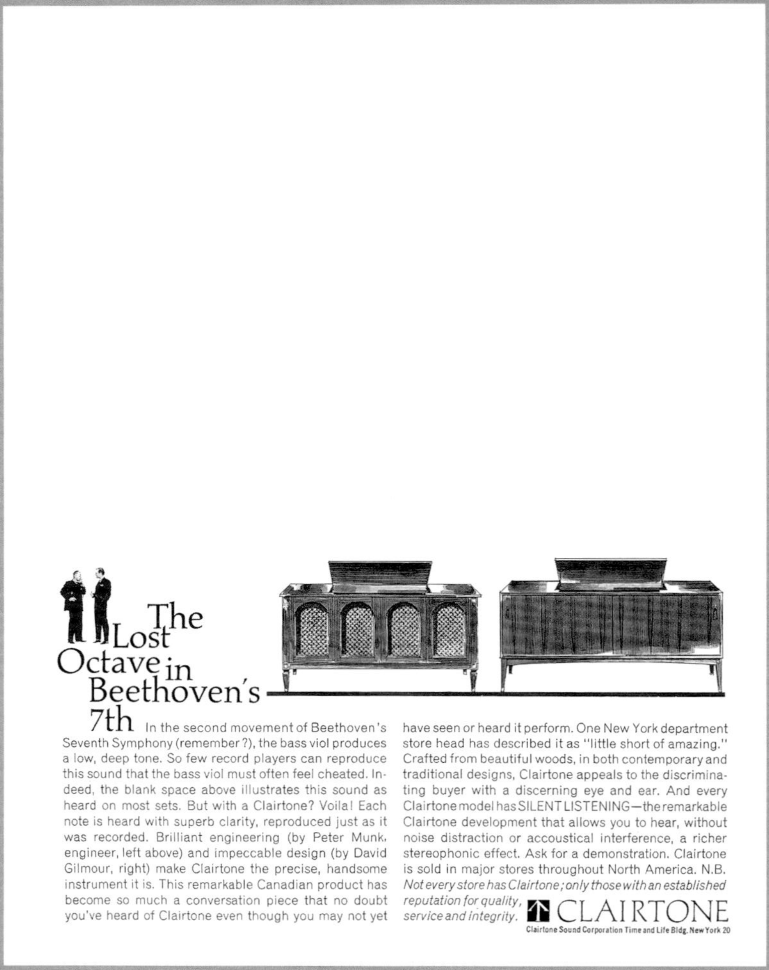

A BIG PRICE FOR EMPTY SPACE

In 1961, Dalton K. Camp and Associates won an award from the Association of Canadian Advertisers for its groundbreaking Clairtone marketing campaign. The judges called the campaign "superb," "quiet," and "convincing." More important, Camp's series of full-page advertisements, which ran in major newspapers and magazines in Canada and the U.S., created an instantly recognizable brand image for Clairtone. The agency's team included Bill Kettlewell as art director and Constance Dinning as copy director. "Peter and David were [initially] unhappy about the Beethoven ad because three-quarters of it was blank—they thought it was a big price for an empty space," recalled Arthur Dollery, a designer for the agency. "But once I explained that the empty space represented the missing note that could only be heard on a Clairtone because the sound was so great, they understood the rationale, and they loved it."

"Wow" is like nowhere, man
The only "wow" Oscar Peterson wants to hear is from audiences all over the world who dig him. There is another kind of "wow" (a mechanical noise) that comes from many of today's record players. No one wants to hear that. Clairtone stereophonic high fidelity sets, however, are designed to be completely free from "wow" and other forms of distortion. You hear no spurious sounds, no false resonances or coloration. All you hear is music. Peter Munk (engineer—left) and David Gilmour (designer—right) make it possible for you to hear Oscar Peterson or any other artist (train whistles, for that matter, if that's your kick) with superb clarity. The engineering required to bring this off is handsomely housed in Mr. Gilmour's impeccably designed cabinets. Clairtone's performance has been hailed at home and abroad. In the United States, the reaction has brought a new appreciation of Canadian engineering and design skills. Clairtone has led the way in Canada's drive for design recognition and acceptance abroad. You'll find Clairtone now in fine stores from Montreal to Monterey, from Halifax to Honolulu. It's a Canadian success story. People who want perfect sound coming from a beautiful cabinet unerringly choose Clairtone. We suggest that you see this remarkable product the next time you are out shopping. See them all, contemporary and traditional models, priced from $399.00. (They're a little more in the United States.) N.B. Not every store has Clairtone; only those with the highest reputation for service, quality and integrity.
CLAIRTONE

CARL DAIR

CARL DAIR (1912–1967)
Dair designed Canada's first original typeface, called "Cartier" in honour of Canada's centenary. His *Design with Type*, an introduction to typography first published in 1952, is still considered a seminal text. To replace Clairtone's rather conventional logo (an old-fashioned crown), Dair designed a swirling, linear pattern that suggests a sound wave. "The pattern originally derives from the photographic track of a pinpoint light suspended from a slowly-turning mobile, a popular graphic device used in the Bauhaus era," according to his colleague Frank Davies. Dair set the corporate name in Palatino, a modern typeface with a hint of classicism, designed by Hermann Zapf of Germany.

ABOVE Dair's sound-wave logo on a promotional record cover.

RIGHT Clairtone's trade booth at the Better Living Show at Toronto's Canadian National Exhibition in the early 1960s. One of the first two-storey trade booths in Canada, it was designed by Hugh Spencer.

CLAIRTONE
RESEARCH AND DEVELOPMENT OF BETTER STEREOPHONIC, HIGH FIDELITY AND ELECTRONIC PRODUCTS
CLAIRTONE

Clairtone's "ultra-modern" plant and head office opened in July 1962 in Rexdale, Ontario, just off the Trans-Canada Highway. With a total floor space of 40,000 square feet, the facility reflected the company's growing ambitions. Its flat roof and large glass windows announced its industrial purpose, as did Hugh Spencer's bold signage.

CHRIS YANEFF

CHRIS YANEFF (1928–2004)
Yaneff graduated from the graphic design program at the Ontario College of Art, and from the school of advertising and marketing at the University of Toronto. In 1956, he opened Chris Yaneff Limited, a public relations and advertising firm that he headed for more than fifty years. Yaneff designed contemporary trademarks and packaging for well-known Canadian corporations: Conklin Shows, Blue Mountain Pottery, The Beer Store, and Windsor Salt were just a few of his clients. He worked for Clairtone on a monthly retainer, designing brochures and annual reports (the 1960 annual report shown here, for example) and writing lively press releases (including this one from 1962).

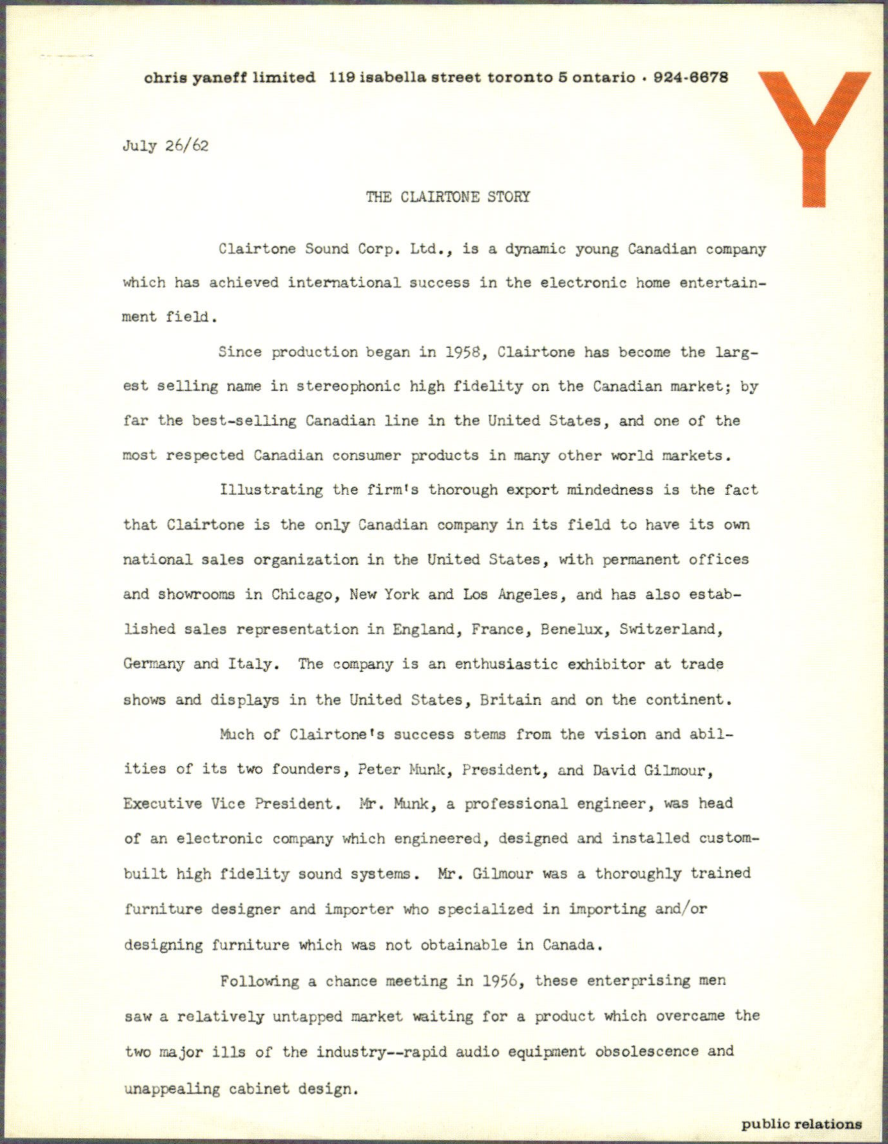

chris yaneff limited 119 isabella street toronto 5 ontario · 924-6678 Y

July 26/62

THE CLAIRTONE STORY

Clairtone Sound Corp. Ltd., is a dynamic young Canadian company which has achieved international success in the electronic home entertainment field.

Since production began in 1958, Clairtone has become the largest selling name in stereophonic high fidelity on the Canadian market; by far the best-selling Canadian line in the United States, and one of the most respected Canadian consumer products in many other world markets.

Illustrating the firm's thorough export mindedness is the fact that Clairtone is the only Canadian company in its field to have its own national sales organization in the United States, with permanent offices and showrooms in Chicago, New York and Los Angeles, and has also established sales representation in England, France, Benelux, Switzerland, Germany and Italy. The company is an enthusiastic exhibitor at trade shows and displays in the United States, Britain and on the continent.

Much of Clairtone's success stems from the vision and abilities of its two founders, Peter Munk, President, and David Gilmour, Executive Vice President. Mr. Munk, a professional engineer, was head of an electronic company which engineered, designed and installed custom-built high fidelity sound systems. Mr. Gilmour was a thoroughly trained furniture designer and importer who specialized in importing and/or designing furniture which was not obtainable in Canada.

Following a chance meeting in 1956, these enterprising men saw a relatively untapped market waiting for a product which overcame the two major ills of the industry--rapid audio equipment obsolescence and unappealing cabinet design.

public relations

For Clairtone's first annual report, Yaneff created a clean and simple design to suit the company's young, modern image. Textured paper in muted colours, and the use of blue to accent the typography, show Yaneff's attention to detail.

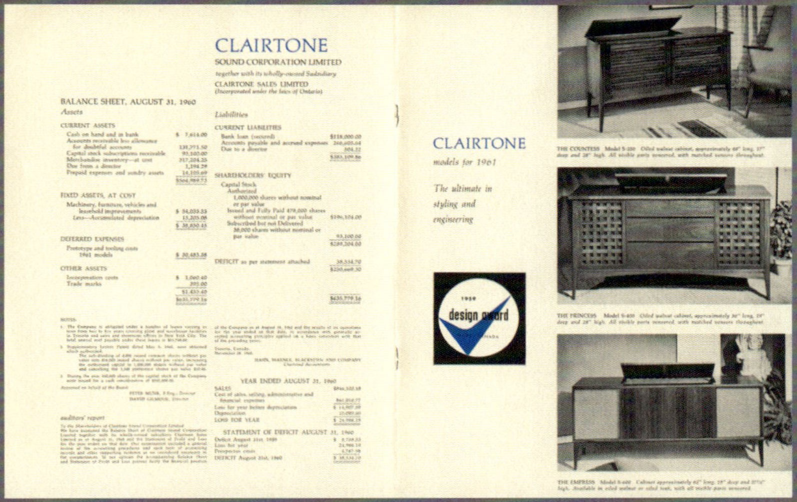

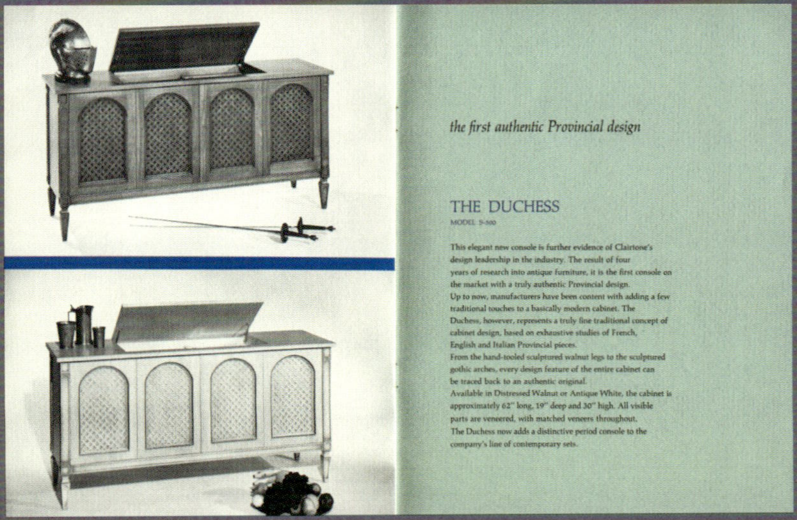

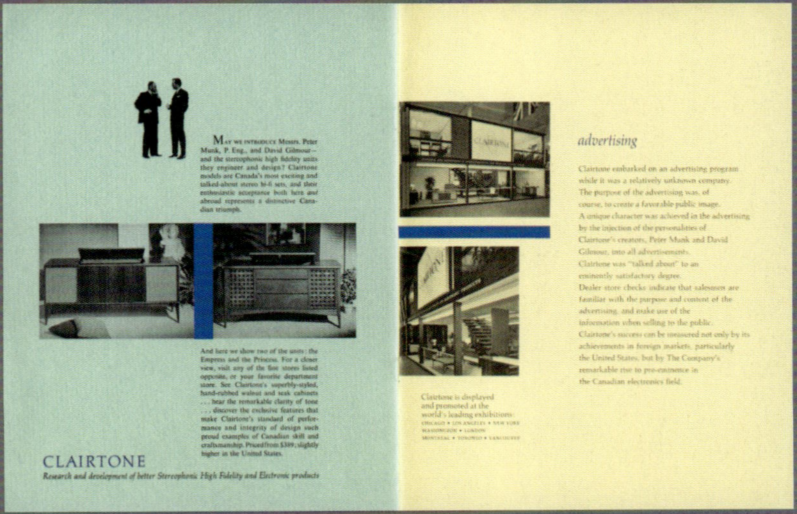

CLAIRTONE
use

LEFT **Visitors crowd around Clairtone's booth in the Better Living Centre at Toronto's Canadian National Exhibition in 1964, eager to experience a "brilliant new concept"—listening to music with headphones.**

ABOVE **Benny Lee, a B-actor from Britain, gives a thumbs up to "Silent Listening" at a Clairtone promotional party at Harrods department store in London. On the far left is Hugh Fraser, heir to the House of Fraser Ltd., the owner of Harrods, with the actress Anthea Wyndham.**

HE DUCHESS"
erco Set by Clairtone
5th Floor

Clairtone's white Duchess stereo on display in the window of a department store.

APPEALING TO WOMEN

Clairtone encouraged dealers to rely on props that suggested privilege and refinement, as shown in this display: a gilded, rococo mirror; a carved candlestick; swag curtains; satin cushions; chinoiserie upholstery; and a plaster bust of Beethoven. "'Try appealing to women,'" Gilmour told dealers, as quoted in a 1961 issue of *Executive* magazine. "'They make all the decisions about furniture.' Hi-fi salesmen were skeptical... But when the dealers found the ladies listening and buying they changed their minds. 'We sell half of them without turning them on,' beamed one major Toronto dealer. 'Styling was the big thing they had to offer in the beginning and nobody else had it.'"

ABOVE **The Empress, model 600-S, introduced in 1960, closely resembles Clairtone's first 100 series. This model boasted a 60-watt, 22-tube amplifier, a Garrard Type A turntable, and EMI speakers. Notice that Clairtone's outdated crown logo is still being used on this early rack card.**

RIGHT **The Princess, model 400-S, also from 1960, was distinguished by its lattice front. Despite its pedestrian looks, it became Clairtone's bestselling stereo. Priced at $399, it was the company's entry-level model, with 30 watts of power and a smaller, five-foot-long cabinet. The blue lines show how the image was originally cropped by Clairtone's art director.**

BRAUN
FM
AM
SW
Volume

LEFT **Clairtone-Braun's T523 portable transistor radio could be inserted in a car by means of a mount bracket.**

RIGHT **The Clairtone-Braun Atelier 2 radio-phonograph combination, with its separate L12 speaker, was notable for its compact design and austere white finish.**

CLAIRTONE-BRAUN

In the mid-1950s, Braun of Germany created a sensation with its highly functional, unornamented consumer electronics designed by Dieter Rams. The Museum of Modern Art, arbiter of good design and good taste, consecrated Braun products by adding them to its permanent collection. In 1961, Clairtone aligned itself with Braun, creating a new division to distribute its transistor radios and stereo components in North America. Today, Braun is a household name; nevertheless, Stan Freeman, who headed the Clairtone-Braun division, remembered that "few people had heard of Braun back then and it was a hard sell." Citing sluggish sales, Braun cancelled its contract with Clairtone in 1964.

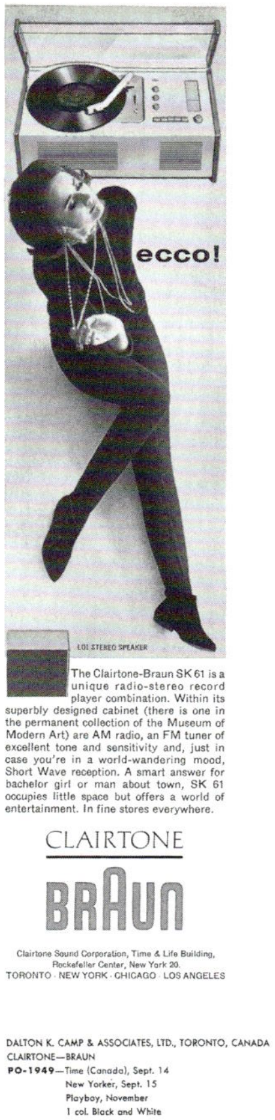

ABOVE **Clairtone-Braun's T221 portable transistor radio, with its purse-shape and vinyl casing, was marketed to women.**

HUGH SPENCER

HUGH SPENCER (1928–1982)
Spencer graduated from the Slade School of Fine Art in London; he then worked in the design department at Granada Television in Manchester and at the BBC in London. After immigrating to Toronto in 1956 he worked at Robert Lawrence Productions, designing sets for the CBC. As early as 1958 Spencer was one of Clairtone's design consultants. He developed a "total design" strategy for the company, using the new Letraset screen system to standardize graphics and signage.

Spencer was as demanding as he was thorough, insisting that the company implement the new look on the faceplates of its chassis, in its displays, and in the corporate logo. Most notably, he designed Project G, his only stereo for Clairtone. At first Spencer worked from his apartment, but after the success of Project G, he was given an office at Clairtone and made director of product development and design.

Spencer left Clairtone in late 1964 and opened a studio called Opus International, where he designed his own furniture line and exhibits for Expo '67. His apartment for Moshe Safdie's Habitat 67 included a one-off stereo console made of concrete to complement the architecture. Instead of specifying Clairtone components for the stereo, he ordered them from RCA, a sure sign of his break with Peter Munk and David Gilmour. In 1969, he moved to Texas, where he designed directional graphics for hospitals and food-trolley systems for airlines. None of these later designs is as memorable as Project G.

LEFT In place of Clairtone's early sound-wave logo, Spencer introduced an upward-pointing arrow based on an ancient runic symbol signifying leadership and fidelity. Bold and asymmetrical, the new logo was embraced by Clairtone. It was "a visual method of communicating an ideal," according to the company.

RIGHT Audaciously, Clairtone's 1961 annual report was produced as an expensive hardcover. The design is striking: transparent vellum juxtaposed on thick, fibrous paper; exquisitely simple graphic charts; and fine renderings of the cabinets.

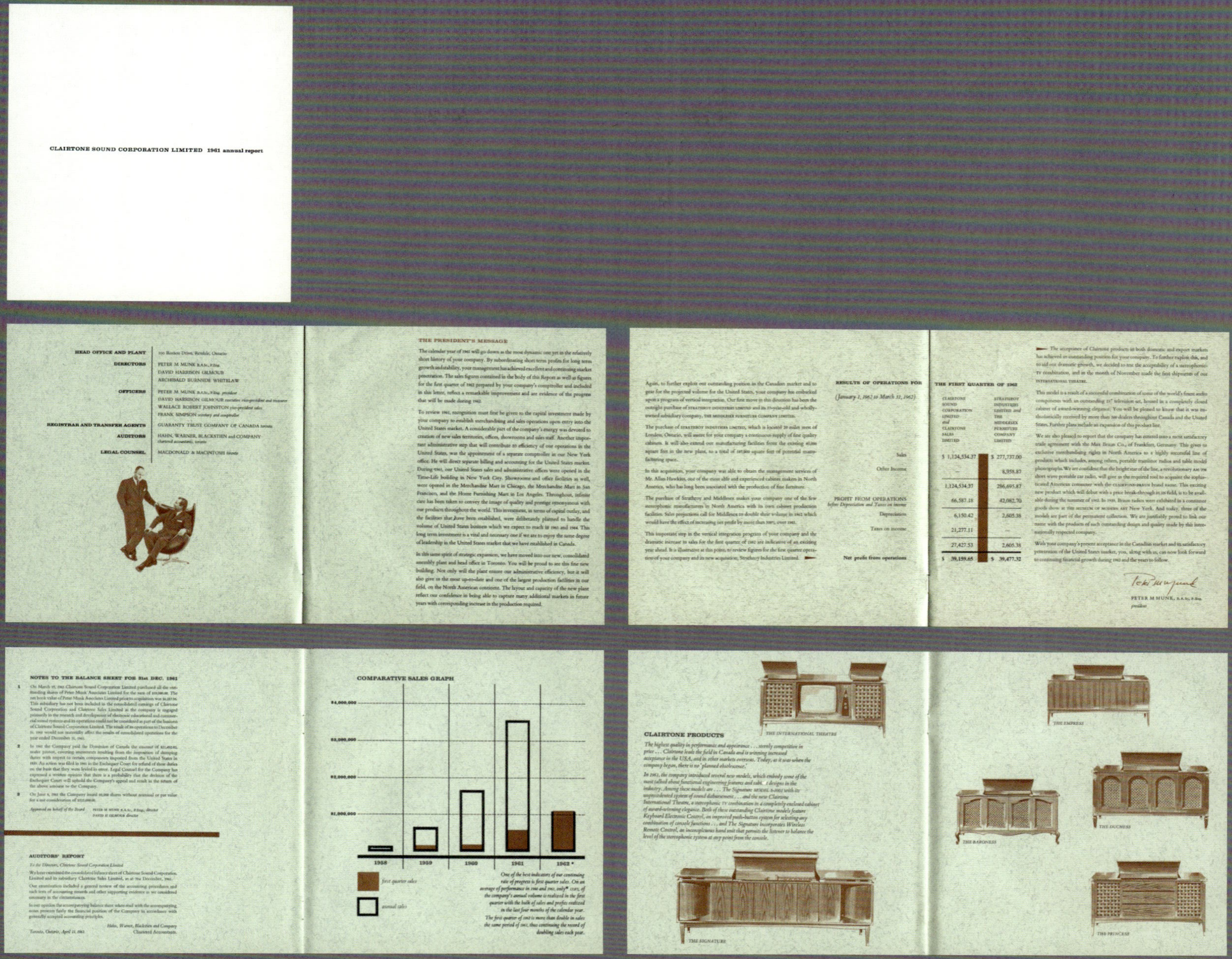
CLAIRTONE SOUND CORPORATION LIMITED 1961 annual report
HEAD OFFICE AND PLANT
DIRECTORS
OFFICERS
REGISTRAR AND TRANSFER AGENTS
AUDITORS
LEGAL COUNSEL
THE PRESIDENT'S MESSAGE
RESULTS OF OPERATIONS FOR THE FIRST QUARTER OF 1962
NOTES TO THE BALANCE SHEET FOR 31st DEC. 1961
AUDITORS' REPORT
COMPARATIVE SALES GRAPH
CLAIRTONE PRODUCTS

stereo fm
phono
tape
tv
mono
reverse
afc
CLAIRTONE
solid state T7
stereo
fm
stereo
earphones
low level
fidelity
FM
88
90
92
94
96
AM
55
60
65
70
80
volume
balance
bass

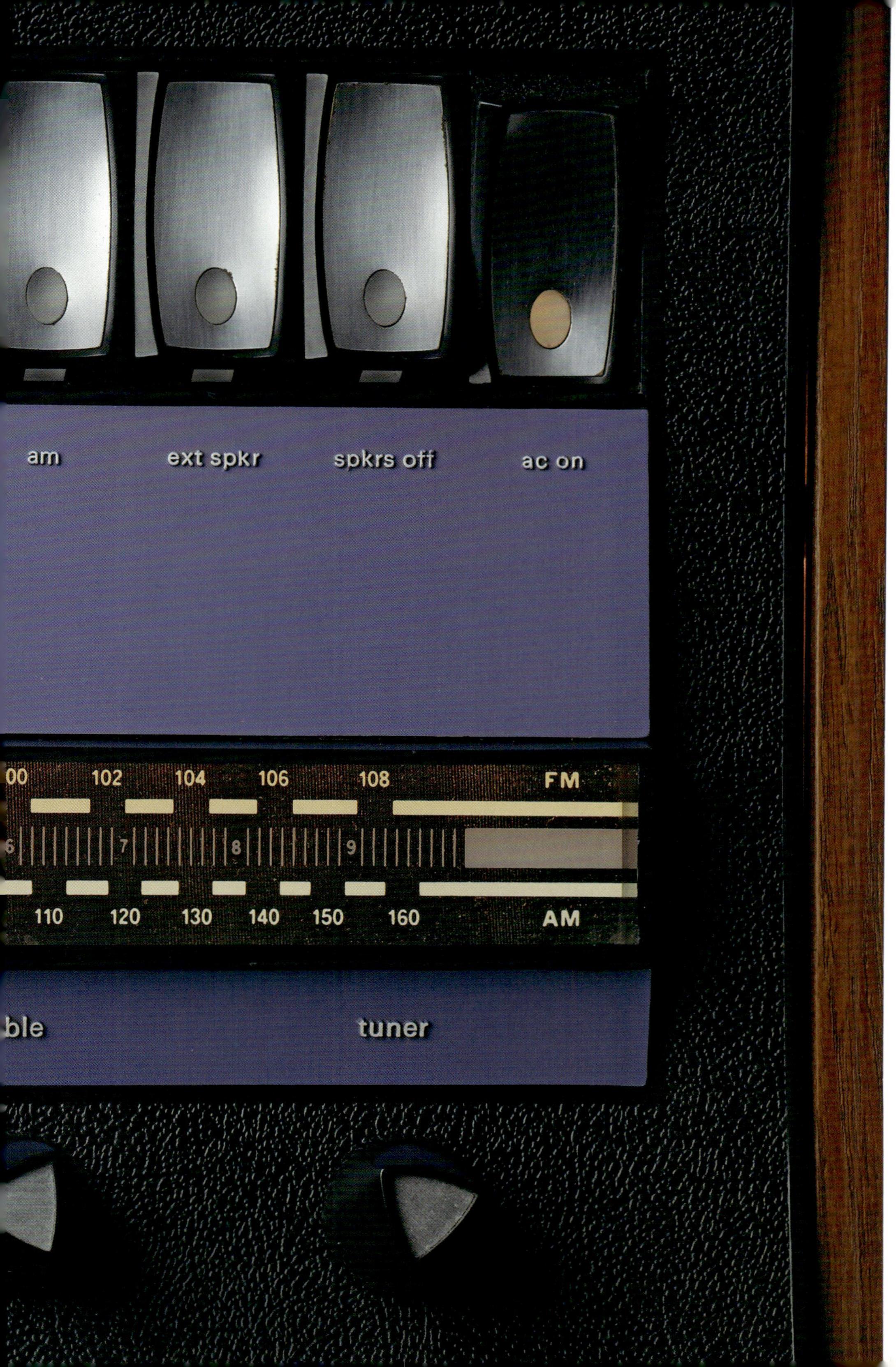

Faceplate of a fully transistorized ("solid state") T7 chassis, designed by Hugh Spencer with Morley Markson, and made for Clairtone by General Instrument of Canada. Spencer designed many of the faceplates for Clairtone, using minimal, modern graphics, a distinctive purple background, triangular knobs, and "rocker" keys.

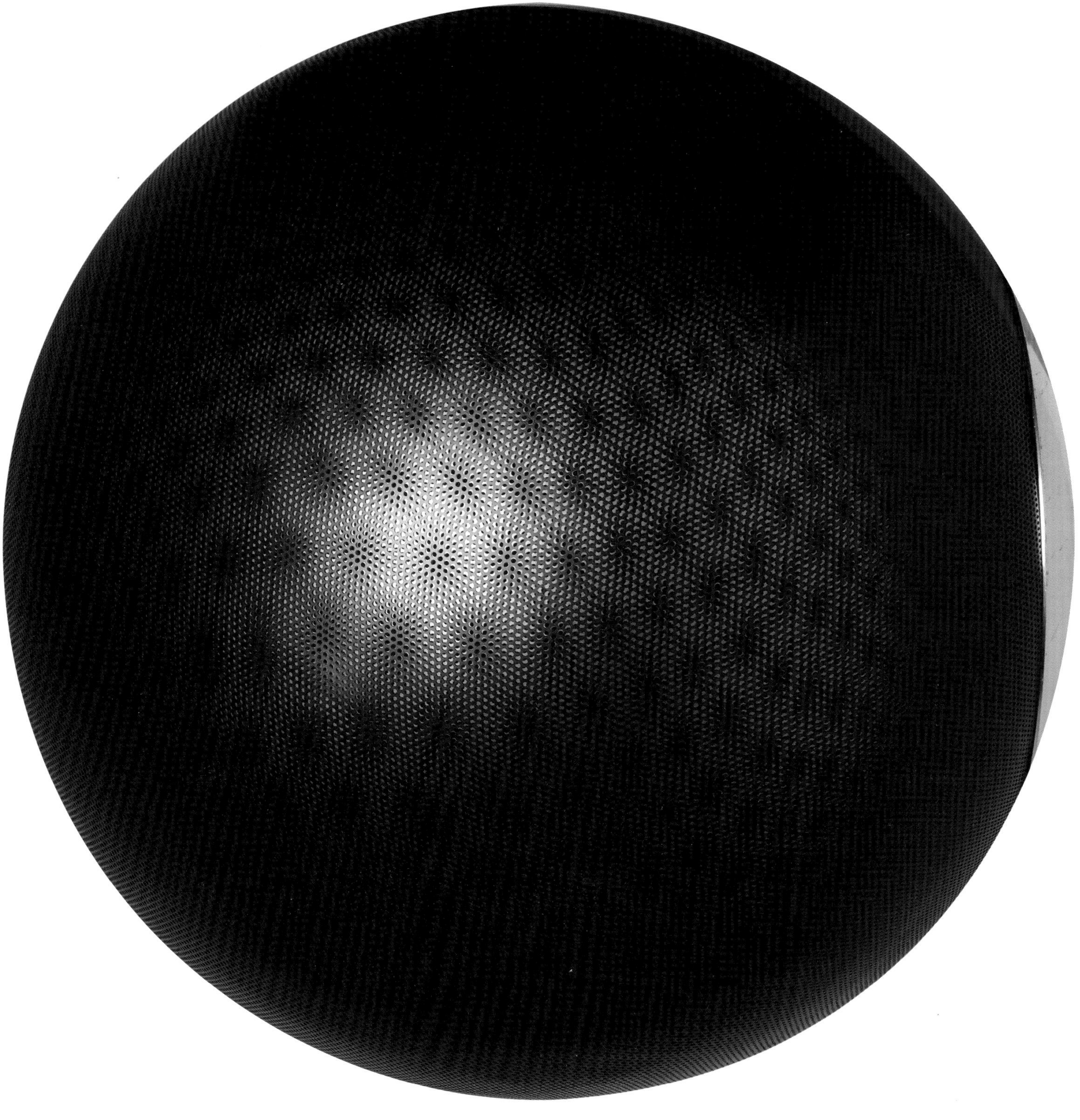

"Clairtone's Project G represents a unique design statement in Canadian history. The choice of materials and its modernist form have made it the iconic Canadian design it is today."

Glenn Pushelberg

INTERIOR DESIGNER, CO-FOUNDER, YABU PUSHELBERG

Project G's 18-inch spun aluminum "sound globes" weren't just a matter of style; they also rotated 340 degrees and, according to company literature, "provided a true sound cone projection, rather than flat layers of sound-wave."

PROJECT G

Introduced at the National Furniture Show in Chicago in January 1964, the Project G is Clairtone's most-celebrated product. "What is so remarkable about Project G," to quote Daniel Leblanc, a sound engineer who collects Clairtone electronics, "is you can't tell which direction the sound comes from; it fills the room like an orchestra. It still sounds that way today." Clairtone produced fewer than 500 Project Gs.

Initial designs for Project G called for a wood base, but when the finished product proved heavier than expected, the designers introduced a metal base on casters, a feature typically used in modern furniture rather than hi-fi sets. To produce the metal carriage, Clairtone contacted Supreme Aluminum Industries in Scarborough, Ontario, a manufacturer of pots and pans. "When we approached Supreme we discovered that they weren't very enthusiastic about interrupting a production run of 100,000 saucepan lids to get into the hi-fi business," said Hugh Spencer. "So we invited their executives up to the plant and Munk gave them his 'Forward with Canada' speech. They agreed."

The first-generation Project G includes the TC4 chassis, manufactured for Clairtone by General Instrument of Canada, an ELAC Miracord 10H turntable, and, inside the spun aluminum "sound globes," a Wharfedale speaker. The TC4 chassis, above, was designed by Hugh Spencer with Clairtone's engineers, and is the only chassis with the name "Project G" on its faceplate.

SEPTEMBER
OF MY YEARS
FRANK SINATRA
Arranged and Conducted by
Gordon Jenkins
reprise
F 1014
SIDE
1
MONO
ELAC
Miracord 10H
MADE IN W. GERMANY · PAT. PEND.
7"
10"
12"
STOP
CLAIRTONE

GARRARD
LABORATORY SERIES AUTOMATIC TRANSCRIPTION TURNTABLE - LAB 80

In 1965, when its contract with General Instrument ended, Clairtone began converting more than 300 Project G cabinets to accommodate the company's new T10 chassis, designed by Al Faux and manufactured in-house. In contrast to the TC4 seen on page 68, the T10 fits into the cabinet horizontally. In other changes, this later model of the G includes a Garrard Lab Series turntable, and the cabinet's interior is partly lined in pink felt (rather than red velvet). According to a 1965 inter-office memo from David Gilmour, the rosewood used for some of the cabinets was too red and the finish had a "plastic appearance" that "was not up to the standard of Danish designed furniture"; he asked that they be refinished.

THE DESIGNER'S NOTEBOOK

As Hugh Spencer's sketches published in the March 1964 issue of *Canadian Interiors* reveal, there was considerable experimentation in Project G's design. Spencer first challenged Clairtone's engineers: "I want you to imagine that you are visitors from Mars and that you have never seen a Canadian living room, let alone a hi-fi set. What are the features that, regardless of design considerations, you would like to see incorporated in a new hi-fi set?" For six months, Spencer worked closely with Clairtone's senior engineer, John Magyar. "They were inseparable, like twins," remembered Anna Spencer, Hugh's wife and an employee in Clairtone's P.R. department. The drawings show that Spencer wanted the speakers to be fibreglass rather than aluminum. He imagined them in white, in black, and in his favourite colour, purple, which he used on Clairtone's faceplates and for exhibits. Remarkably, Spencer was colour-blind. "For Hugh, purple was dark grey," reported Tony Mann, a friend and colleague. "He always used purple because someone told him it was grey." Even if Spencer's use of purple was unintentional, it became his trademark colour. Explained Anna Spencer: "He believed purple was a very architectural colour… He always said that he was a designer who believed in 'a touch of vulgarity'—it pulled the eye and made a statement."

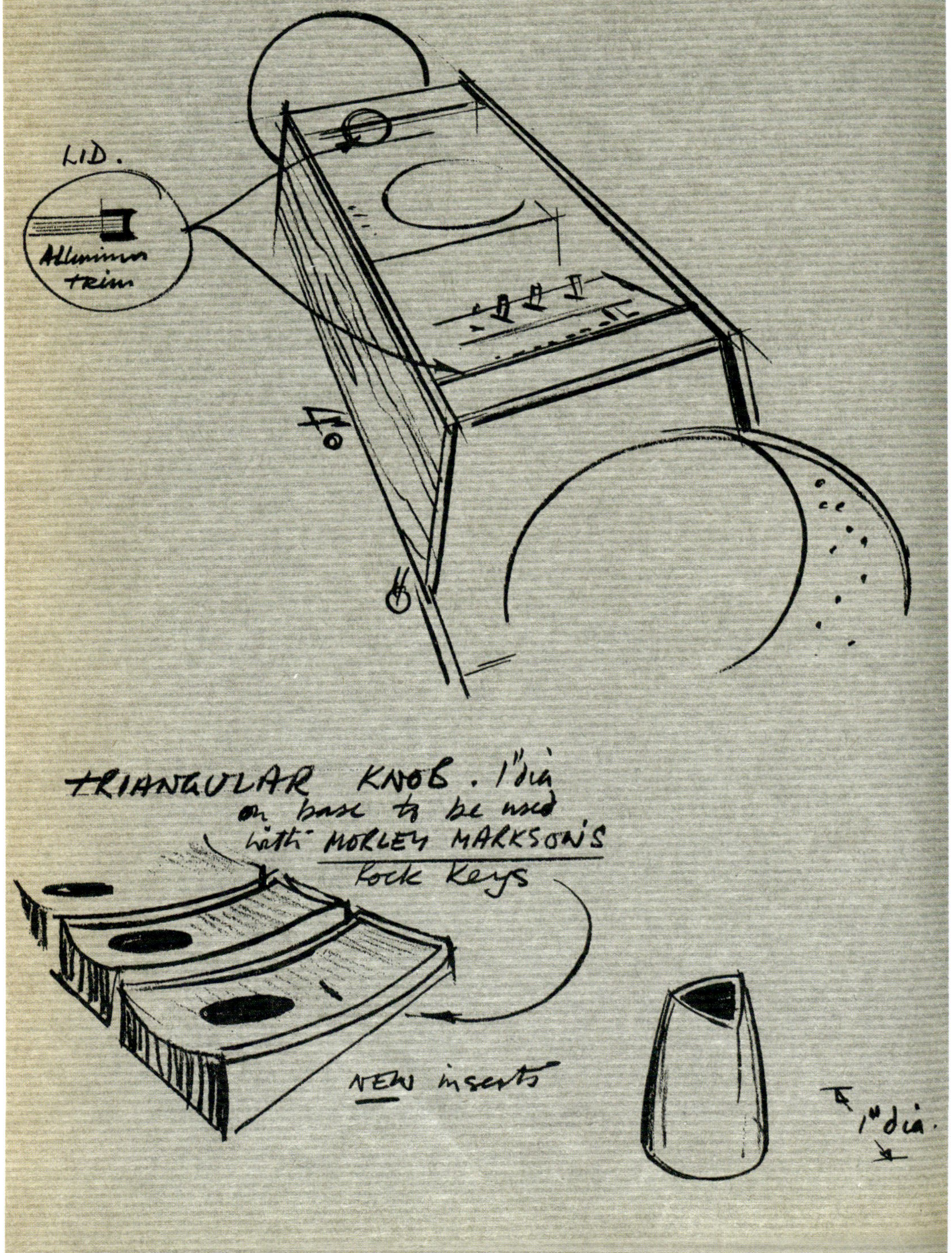

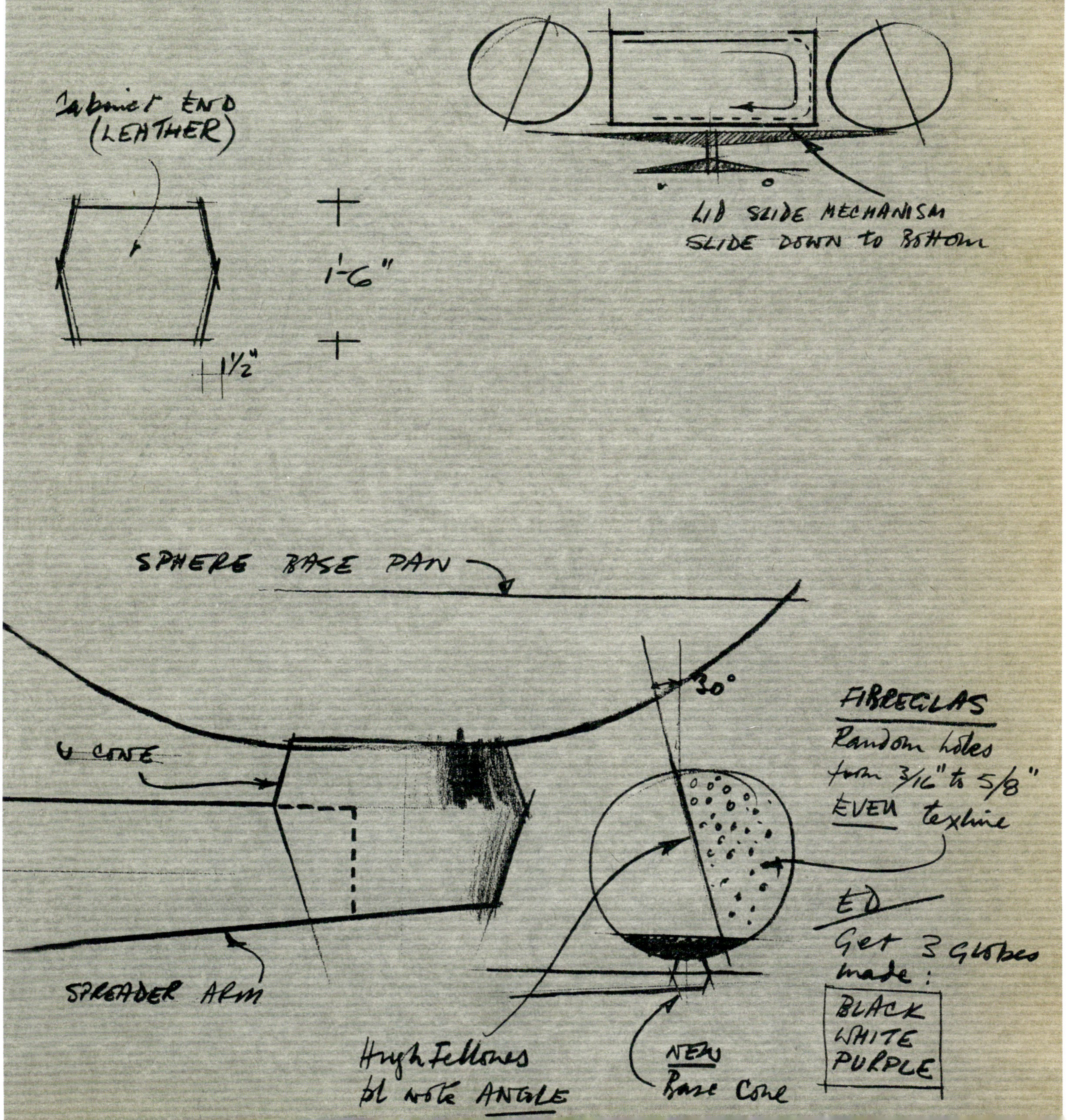

Cabinet END
(LEATHER)
1'-6"
1½"
LID SLIDE MECHANISM
SLIDE DOWN to BOTTOM
SPHERE BASE PAN
30°
FIBREGLAS
Random holes
from 3/16" to 5/8"
EVEN texture
ED
Get 3 Globes
made:
BLACK
WHITE
PURPLE
CONE
SPREADER ARM
Hugh Fellowes
pl note ANGLE
NEW
Base Cone

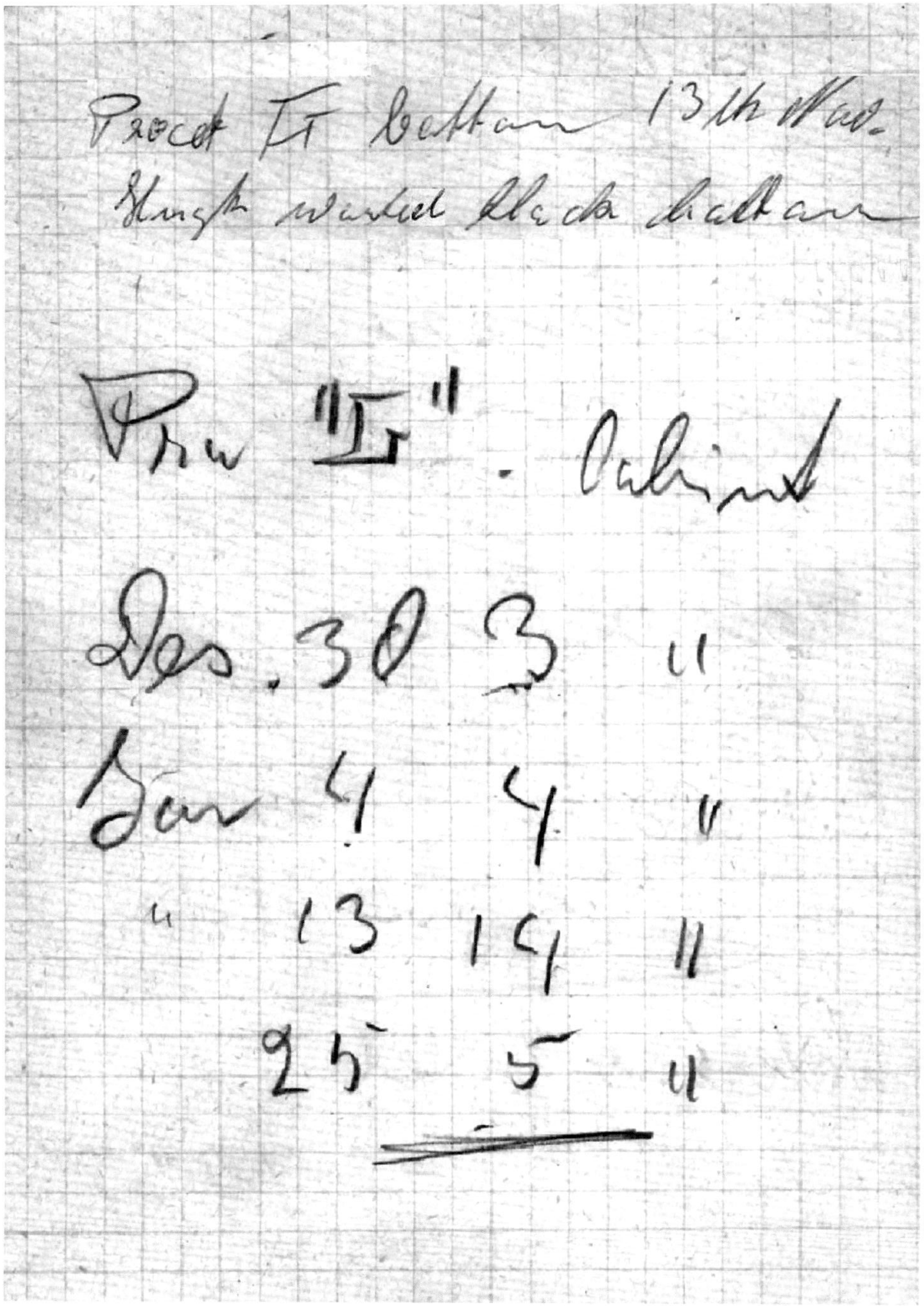

RIGHT **Cabinetmaker George Weil's notebook entry, dated November 13, 1962, reveals early guidelines for the production of Project G prototypes. "Hugh wanted black bottom," noted Weil, who was ordered to make 26 cabinets between December 30 and January 25.**

FAR RIGHT **Peter Munk on the Project G assembly line in Rexdale, 1965.**

NOW FOR MY NEXT TRICK

By early 1964, when Project G was introduced, Clairtone's U.S. sales had begun to stall. Peter Munk and David Gilmour attributed this to the fact that Clairtone was relatively unknown in the U.S. As Munk noted in an interview with *Business Week*, "So far as people in the States are concerned, Canada means wheat, fish, and endless silent forests. Nobody thinks of us as highly skilled manufacturers." Project G was going to change all of that; so important was its launch that Munk and Gilmour spent months personally presenting their new stereo to every major retail executive in North America and Europe. For a time, only a single working prototype of Project G existed, often forcing Munk and Gilmour to sell the model from snapshots alone. Sometimes the prototype malfunctioned. In an interview, Gilmour recalled such a time: "Mr. Morton May of May Department Stores in St. Louis filled the auditorium with his best customers. He gave me a velvet cover for Project G. I stood up before a few hundred people and explained this moment of breakthrough in acoustics, and the sound and the joy and the pleasure. I took off the velvet lid, turned on the sound that I had preset, this rich sound—and like an atomic cloud, a puff of smoke came up! I took the velvet cloth and quickly threw it back on. Then I said, 'And now for my next trick…'"

FAR LEFT **Peter Munk shows Project G to a group of executives.**

ABOVE **David Gilmour demonstrates Project G in Copenhagen, with Svein Grundt from Dube Electronics, one of Clairtone's European distributors.**

LEFT **Gilmour presents Project G to Hugh Fraser of Harrods.**

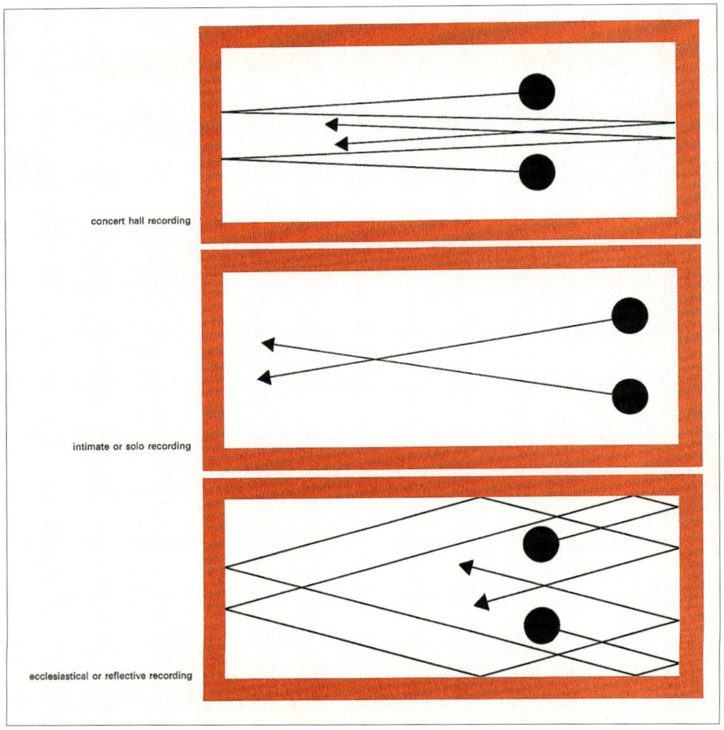

THE ART OF PROJECT G

Clairtone put almost as much energy into the creation of the first Project G brochure as it did into Project G itself. Hugh Spencer designed it and hired Irving Penn to photograph the stereo. Spencer's wife, Anna, recalled the photo shoot: "We took the prototype and strapped it on top of our Pontiac wagon. Penn had a very modest and unobtrusive studio in midtown Manhattan. There was no parking and the shipping elevator was small. We had the G swathed in shipping blankets, up-ended on one globe speaker, and Hugh and I and Penn's assistant literally held it up in the air whilst the elevator went up to the studio." Spencer's cover design is discreet, and Penn's extraordinary photograph takes centre stage. By emphasizing the black globe and minimizing the cabinetry, Penn's vertical composition makes Project G into a tall sculpture of pure sound. Spencer's bold illustrations, executed by Art Associates in Toronto, echo the sculptural qualities of Project G. Note to collectors: the chassis in the prototype is the XB7 from Clairtone's basic line; the TC4 was not yet available.

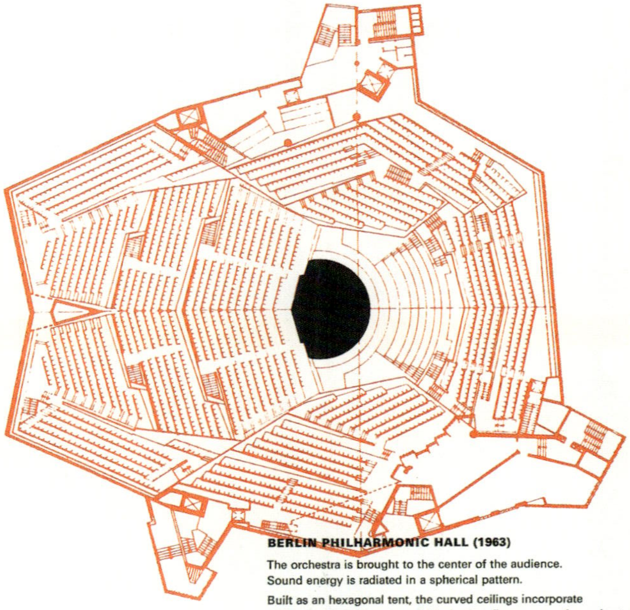

BERLIN PHILHARMONIC HALL (1963)

The orchestra is brought to the center of the audience.
Sound energy is radiated in a spherical pattern.

Built as an hexagonal tent, the curved ceilings incorporate 136 pyramidal reflectors and 10 plastic reflectors over the orchestra.
The walls are of teak and exposed concrete depending upon the degree of reflection or absorption required.

5 application of concert hall acoustics

Today's architects are releasing the orchestra from the confines of the concert hall proscenium—so Project G's* sound globes* free the speaker from the confines of the conventional rectangular box enclosure and sound energy is radiated spherically.

6 significant advances in engineering

The sound globes* are mounted on pressure adjustable nylon bearings supported on cantilevered arms outside the cabinet.
Mechanical feedback from speaker vibration is eliminated.

The cantilevered arms become mechanical filters.

A special circuitry has been developed to measure distortion in the cone movement.

This circuitry automatically adjusts the output of the power amplifier to obtain minimum distortion of the sound.

The background notes were taken from the design engineer's personal lab log.

*trademark

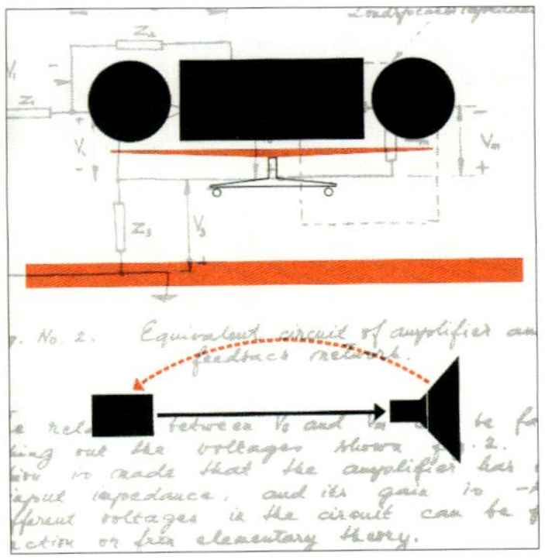

7 total design

With the development and use of the all-transistorized chassis, no heat is generated, no ventilation is needed and the unit is finished on all sides.

Project G was designed as a totally enclosed sound unit where form and function are related to create a shape comparable to today's architectural furniture design.

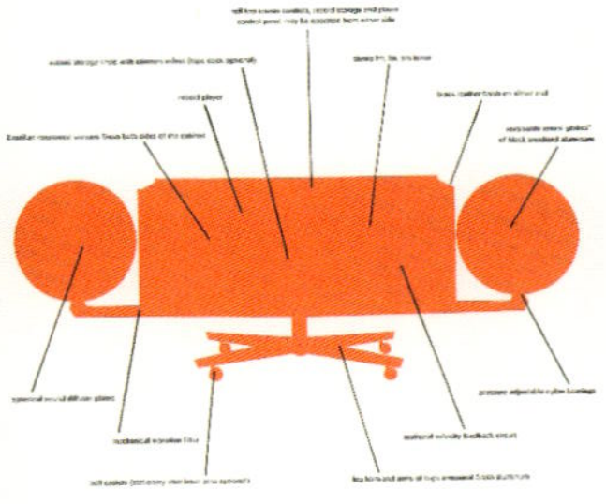

FAR LEFT **Project G and Clairtone's first in-house model, Anna Gilmour.**

ABOVE **Behind Peter Munk, Clairtone executives, from left to right: Leslie Berenyi, Eric Smith (obscured), David Gilmour, and Zigmund Hahn. Dalton Camp's ad campaign is on the far wall.**

AN ICONIC PROP

In the canon of good design, chairs hold the place of honour. It is not surprising that Clairtone promoted the design of its stereos by juxtaposing them with modern furniture. But why choose Arne Jacobsen's 1956 Egg chair so often? Gilmour bought one in the late 1950s when he ran Scantrade, an import business; he then moved the Egg chair into his Clairtone office. "To me it was one of the greatest designs of furniture of its time," said Gilmour. "It became an iconic prop for us; it just seemed to have a personality of its own that went with the technology."

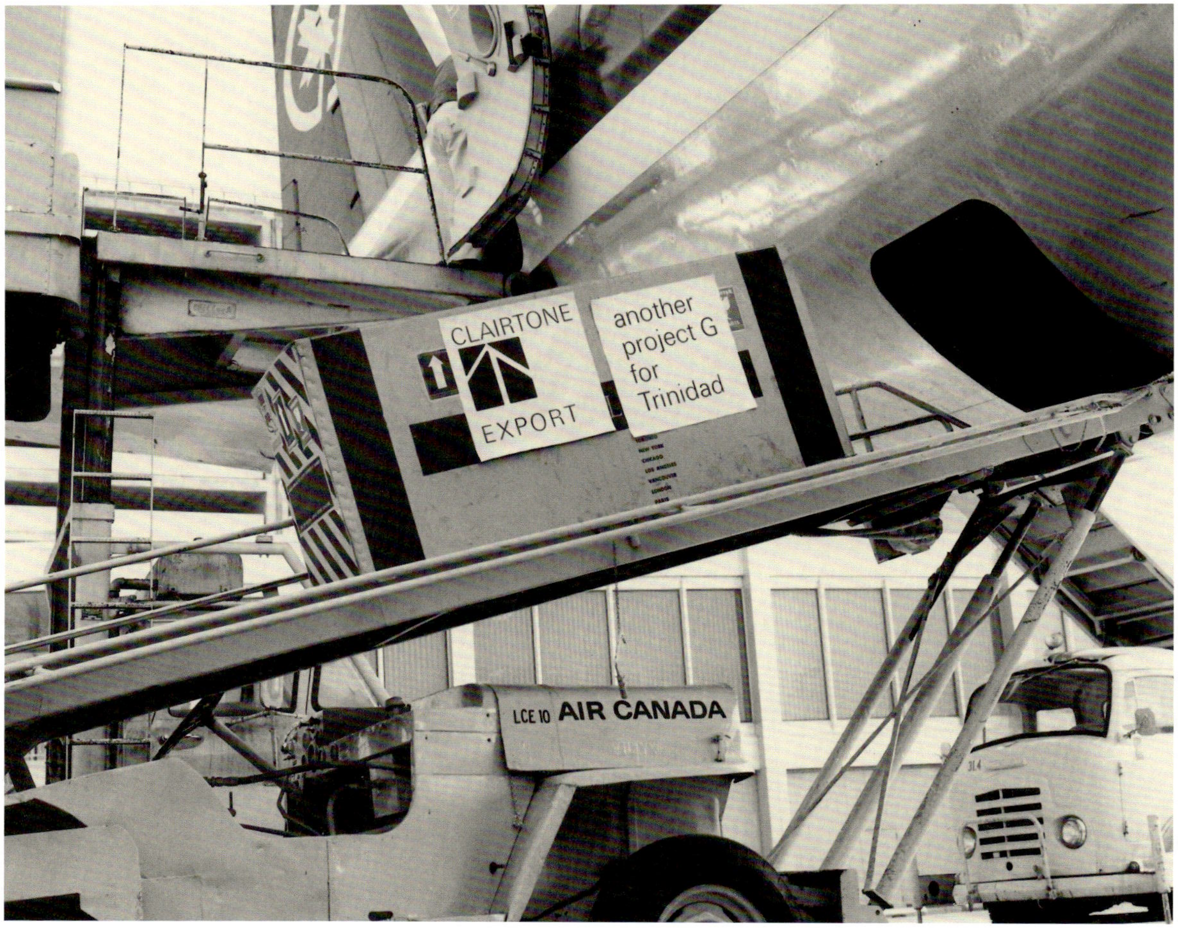

RIGHT **A Project G being loaded onto an Air Canada plane.**

FAR RIGHT **Oscar Peterson looks on as a Clairtone employee fits a TC4 chassis into a Project G, at Clairtone's factory in Rexdale, Ontario.**

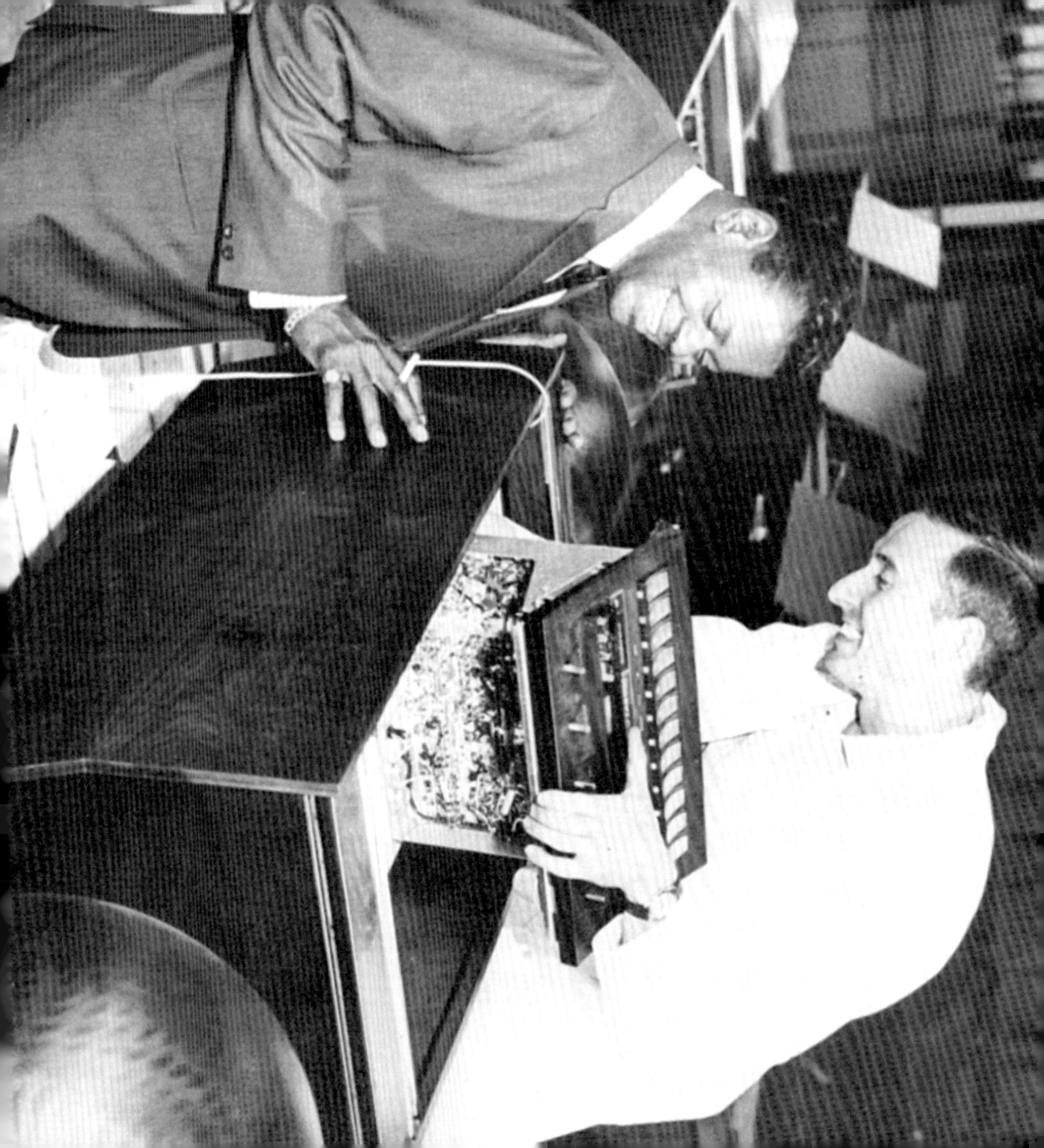

“The design of Project G is as ebullient and optimistic as the zeitgeist of the ’60s. It is no surprise that it has endured in our imagination as an icon of Canadian design.”

Shirley Blumberg

ARCHITECT, FOUNDING PARTNER, KPMB

At the annual Better Living Show, Clairtone invariably created excitement at its booth. In 1965, it presented Katyha, the corporate muse who twisted and turned to the latest discotheque music. The Clairtone newsletter reported that some 3,000,000 people passed through the show that year, and most of them dropped by to see “yé yé girl” Katyha, whose full name (the less exotic Katyha MacConnell) was rarely used.

JUNGLE
DELIGHT

THE POWER OF HOLLYWOOD

Knowing the power of celebrity endorsements, Clairtone hired a product-placement agency in Los Angeles. First came *Marriage on the Rocks* (Warner Brothers, 1965), a film starring Frank Sinatra and Dean Martin as playboys whose props include a Malibu beach house and the seductive sound of a Project G. More product placements followed: A Project G is playing when Frankie Avalon sings to Tuesday Weld in *I'll Take Sweden* (United Artists, 1965), and in *The Oscar* (Embassy Pictures,1966), a G is part of Tony Bennett's Hollywood bachelor pad. In Sonny and Cher's *Good Times* (Columbia Pictures, 1967), a Project G serves as a perch for a vase of bird of paradise flowers.

FAR LEFT Tara Ashton and Frank Sinatra in publicity shots from Marriage on the Rocks.

TOP Tony Bennett and a Project G.

CENTRE Frankie Avalon with a G.

BOTTOM On the set of a Columbia Pictures movie, an unidentified actress and Project G.

PAGE 88 Sonny and Cher on the set of Good Times.

Patrician, finely detailed Mediterranean design in satin walnut. Like all Clairtone consoles features solid state tuner and amplifier of sophisticated design. $499.00

Famous people everywhere applaud Project G. It's the only console in the world that lets you move the sound to suit room acoustics and music. "G" wins design awards too. $1600.00

Contessa is fashion conscious in a very elegant way and features the Clairtone-exclusive Automatic Fm Selector for touch key tuning of fm/stereo stations. $789.00

Listen to Sinatra on Clairtone stereo.

Sinatra does.

See him in Marriage on the Rocks, an A-C Production through Warner Brothers.

Clairtone stereo is one of the small American miracles that comes along every now and again to make life more pleasant. ☐ When we started Clairtone seven years ago we had four employees, including the President and Executive Vice-President. The first year we sold 350 sets and had the smallest Christmas party in the stereo business. ☐ To cut a long story short, this past summer we sold Clairtone set number 100,000 and are now the world's biggest specialists in package stereo sets. Bar none. ☐ But what we're proudest of is our customers, because they include a long list of people who are famous for knowing what music should sound like. For instance, Frank Sinatra. ☐ If you'd like to hear why people like this choose Clairtone you can go and see and listen to one in virtually any famous department store in North America, or a select number of music houses and appliance shops. Please do.

LISTEN TO SINATRA

As part of its marketing strategy to reach an "elite group" of "trend setters," Clairtone eagerly sought endorsements from talented musicians. Oscar Peterson, the legendary Canadian jazz pianist and an early endorser of Clairtone, was introduced to the company through his friend Eric Smith, a senior sales executive at Clairtone. Peterson in turn introduced Clairtone to other big-name jazz artists, including Carmen McRae and Joe Williams. Clairtone's biggest coup was using Frank Sinatra as the centrepiece of its 1965 campaign. According to newspaper accounts of the time, Sinatra was so taken by Clairtone's stereos that he gave half a dozen to his Rat Pack friends. The ad campaign, which also featured Stan Getz, was created by Group X in Toronto, a small, independent division of McCann-Erickson, with Cubby Marcus and Leo Brouse as copywriters and George Pastic as art director. The ads ran in *Life*, *Vogue*, and Igor Cassini's *Status* magazine, among other publications.

At Clairtone's Rexdale showroom with Project G, from left to right: "Cannonball" Adderley, Carmen McRae, and Joe Williams.

Tredicesima
Triennale di Milano

Tempo libero
Esposizione internazionale delle arti decorative e industriali moderne e dell'architettura moderna

Leisure
International exhibition of modern decorative and industrial arts and of modern architecture

Freizeitgestaltung
Internationale Ausstellung für moderne dekorative und angewandte Kunst und moderne Architektur

Loisirs
Exposition internationale des arts décoratifs et industriels modernes et de l'architecture contemporaine

diploma di medaglia d'argento

Clairtone Sound Corporation

Il Presidente
della Giuria Superiore
Prof. Bruno Molajoli

Il Presidente
della Triennale di Milano
Dino Gentili

CANADIANA

Project G was awarded a silver medal in September 1964 at the 13th Milan Triennale, the international design exhibition. In total, Canada won five medals: two gold and three silver. The Canadian pavilion, shown here, was organized by the National Design Council, which built a country cottage around that year's theme of leisure. The Ottawa-based architectural firm Schoeler and Barkham was hired to design the cabin, a simple, modern box of cedar and glulam, a new load-bearing composite made of Douglas fir. Canadian paintings, Inuit art, a Project G, and a teepee-shaped metal fireplace by Court Noxon were part of the installation. The all-Canadian cabin had a distinctive Scandinavian feel, perhaps because many of the furnishings were designed by Danes who had recently immigrated to Canada: Lotte Bostlund designed the ceramic lighting; Karen Bulow was responsible for the handwoven drapery; and a Canadian company called Scandia Furniture made the folding chairs. To quote the August 1964 issue of *Canadian Interiors*: "Clairtone's Project G looks a little citified and expensive for a cottage. But it is good Canadian design, and, as such, belongs in a design exposition."

LEFT Between 1965 and 1967, Clairtone sponsored the Miss Canada beauty pageant. Miss Canada 1966 was Diane Landry, seen here with a G2 at Clairtone's factory showroom in Rexdale.

RIGHT A 1966 promotional tent card for the G2.

ERA TWO

By the mid-1960s, following the spirit of the time, Clairtone's marketing department moved away from the image of "gracious living" to embrace the Swinging Sixties. Clairtone's first in-house model, Anna Gilmour, was replaced by a provocative Go-Go dancer named Katyha. Project G appeared in *Playboy* magazine and in the influential movie *The Graduate*. Clairtone's new advertising agency, McCann-Erickson, launched a campaign that focused on the company's iconoclastic "sound globes." Other designers came on board: Al Faux, Tony Mann, and Frank Davies.

The year 1966 marked the launch of Clairtone's G2 stereo, a smaller and, at $950, less expensive version of the short-lived Project G. "There was always the intention for a G collection," explained Peter Munk. "The G was a showpiece, a conversation piece; it was meant to play in a concert hall. The G2 was more utilitarian, more useful to the average person." Clairtone's engineers boasted that the G2 was more refined than the G; that despite having less heft and power, its speakers produced more clarity and a wider spectrum of sound. In the words of Tony Mann, one of the G2's designers: "The G was the Thunderbird and we tried for the Ferrari."

While promoting the G2 to audiophiles and modernists, Clairtone maintained its position in the traditional hi-fi market, sponsoring the Miss Canada pageant and continuing to make old-fashioned cabinets to match living-room furniture. As the company's literature explained, Clairtone's product line included "everything from Spanish to Danish to Italian Provincial to Early American."

For Munk and Gilmour, the challenge now was figuring out how to transform Clairtone from a respected, but fringe company to a multinational electronics firm in the same league as Zenith, RCA Victor, and Magnavox. The solution, they decided, was to build a massive, modern factory—a factory that would eventually be built with funding from the government of Nova Scotia.

A G2 prototype, designed in 1965 by Al Faux, shows clearly his original intention to create a minimalist structure: two distinct boxes appear to float above a polished chrome T-shaped base. Cantilevered on either end are the globe speakers. This prototype was quickly replaced by a sturdier, but less ethereal G2 that includes a solid metal support bar between the boxes and the base.

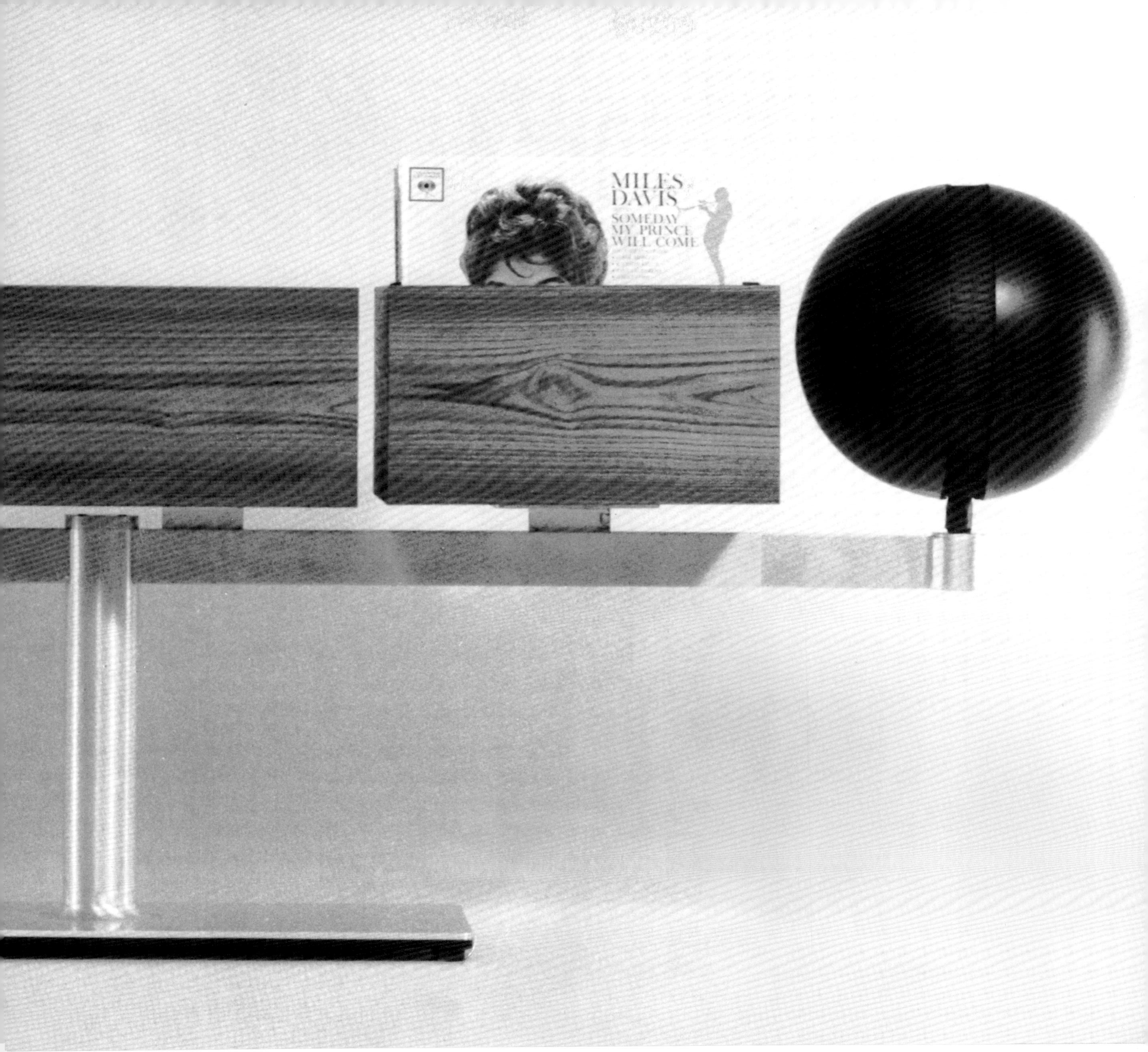
MILES DAVIS
SOMEDAY MY PRINCE WILL COME

LEFT In selling the G2, Clairtone targeted design buffs and interior decorators, encouraging them to experiment with new looks.

RIGHT Shown with the G2 in this marketing photo are two Florence Knoll armchairs and an Eames Aluminum Group chair. To underline the G2's versatility, the detachable globe speakers have been placed on the bookshelf. For a time, a version of the G2 was available without a base, and an additional pair of speakers could be purchased separately.

10GUINEAS
VINES
GRAPES
Green-House Plants
PLANTS

A G2 adds to the eclecticism of this living room designed by Milo Baugham for a client in White Plains, New York. Published in early editions of Ray and Sarah Faulkner's best-selling book, Inside Today's Home, the interior is described as "a striking example of contemporary freedom of taste." High modernism is ignored; instead, African sculpture is juxtaposed with Op art, and the natural wood grain of the sloped ceiling and of the G2 is set off by plush, brightly coloured velvets.

Canada–Design '67

Award of Excellence
Prix d'excellence

A Programme Developed by the National Design Council and Department of Industry to Recognize, Promote and Encourage High Standards of Canadian Product Design

Programme mis en oeuvre par le Conseil national de l'esthétique industrielle et le ministère de l'Industrie en vue d'apprécier, de récompenser et d'encourager des normes d'esthétique de haute valeur dans la conception des produits canadiens

In "New Product Design" category for Originality of Design in

Dans la catégorie des "Nouveaux Produits" pour la conception originale de

Stereo AM/FM Phonograph Console and Components
Electrophone stéréophonique AM/FM et ses pièces

Award of Excellence Presented to
Prix d'excellence accordé à

Clairtone Sound Corp.

J C Parkin
Chairman, National Design Council
Président, Conseil national de l'esthétique industrielle

The Hon. C M Drury
Minister, Department of Industry
Ministre de l'Industrie

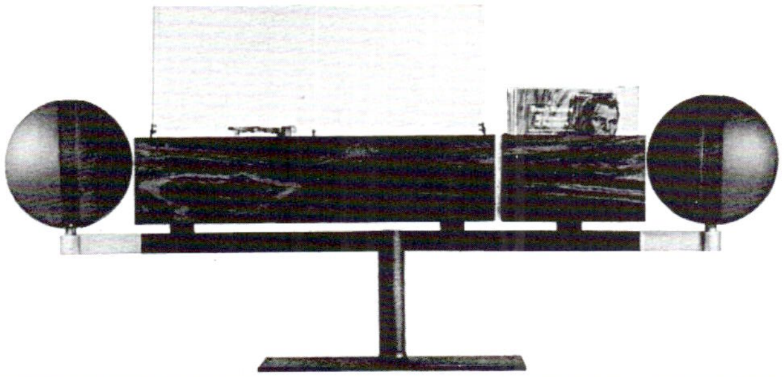

VANGUARD

"Clairtone's cabinet unit defined not only technical innovation but created a true statement piece for lounges, sitting rooms, dens, and studies across North America. In many ways, Clairtone's Project G was a perfect expression of Canada's modern spirit in the run-up to Expo '67 in Montreal."

Tyler Brûlé

JOURNALIST, BRAND CONSULTANT, FOUNDER, WINKREATIVE AND *MONOCLE*

Katyha and an early G2, with Clairtone's T10 chassis and a Garrard turntable.

AL FAUX

AL FAUX (1931–1978)
Faux trained as a furniture designer at Toronto's Ryerson Polytechnic Institute (now Ryerson University) and began his career working for some of Canada's leading industrial designers, including Robin Bush and Lawrie McIntosh. In late 1964 and 1965, as a designer at Hugh Spencer's firm, Opus International, Faux (pronounced "fox") began to work on the G2. After a short time, he left Opus to form Design Collaborative with Tony Mann and continued to design the G2. Faux's final G2 is sleeker and less intimidating than Project G. Thanks to advances by Clairtone's engineering team, it's also less fragile. In his other work for Clairtone, Faux designed faceplates for the T10 and T14 chassis used in the G2. Moving away from Spencer's earlier, stylized faceplates, Faux's designs emphasized function above all. In 1968, Faux established his own company specializing in furniture systems: he designed an award-winning drafting table for Norman Wade Co. and, in one major commission, worked on the University of Guelph's library. He died at the age of forty-seven, at the peak of his career.

LEFT Faux's study for a record storage unit that never went into production.

RIGHT Minutes from a January 29, 1965 Clairtone meeting, discussing Opus's proposed G2 model. Listed as present at the meeting: Faux, Michael Chojnacki (V.P., manufacturing), Ed Steger (procurement), Peter Munk, David Gilmour, and Gloria Collinson (marketing and publicity).

FAR RIGHT Sketches and notes taken at a meeting with Indesco Limited, a company Clairtone considered hiring to produce the G2. While fibreglass and spun aluminum are mentioned here as materials for the speakers, the actual G2 speakers are made of cast aluminum.

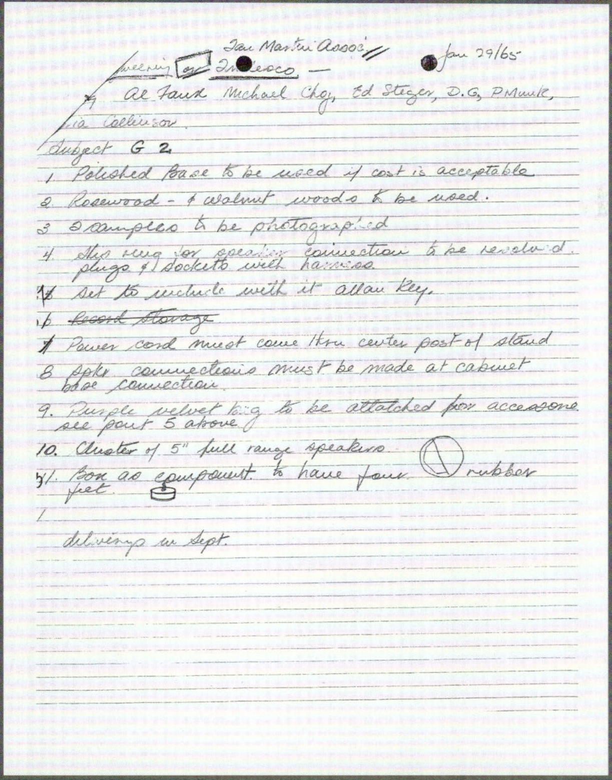
Ian Martin Assoc.
Meeting ~~or~~ Indesco — Jan 29/65
Al Faux Michael Choj, Ed Steger, D.G., P.Munk,
Gia Collinson
Subject G 2.
1. Polished Base to be used if cost is acceptable
2. Rosewood - & walnut woods to be used.
3 Samples to be photographed
4. Mis wig for speaker connection to be resolved. plugs & sockets with harness
5 Set to include with it allan key.
6 ~~Record Storage~~
7 Power cord must come thru center post of stand
8 Spkr. connections must be made at cabinet base connection.
9. Purple velvet bag to be attached for accessories see point 5 above
10. Cluster of 5" full range speakers.
11. Box as component to have four rubber feet.
delivery in Sept.

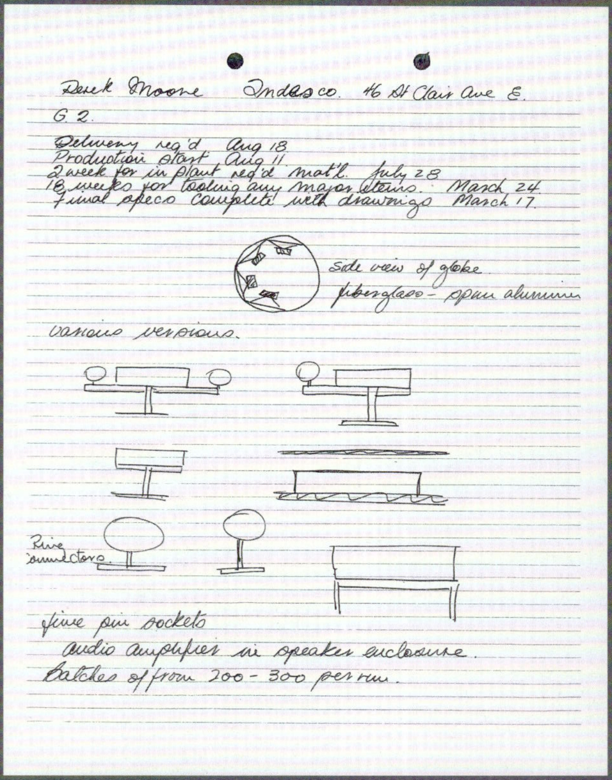
Derek Moore Indesco. 46 St Clair Ave E.
G. 2.
Delivery req'd Aug 18.
Production start Aug 11.
2 week for in plant req'd mat'l. July 28
18 weeks for tooling any major items. March 24
Final specs complete with drawings March 17.
Side view of globe
fibreglass - spun aluminum
various versions.
Five connectors.
five pin sockets
audio amplifier in speaker enclosure.
Batches of from 200 - 300 per run.

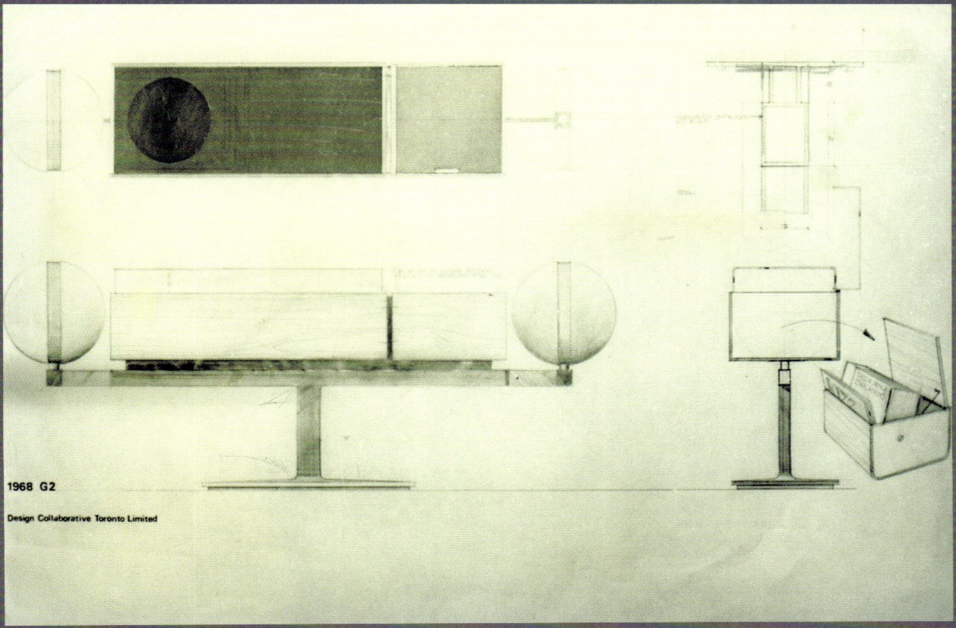

LEFT Faux's drawing shows modifications to stabilize the G2. Changes include an additional support bar between the base and console, and a wood insert connecting the two cabinets. To create the appearance of seamlessness, the new parts are painted black.

G2

The second generation of Clairtone's G series, the G2 of 1966, is more streamlined than Project G. At 78.5 inches, it's nearly five inches shorter than the G. It's also narrower (14.5 inches versus 18.5). In place of Project G's tambour roll-top, the G2's components are protected by a bi-level transparent plastic dust cover. The most noteworthy difference is the G2's modularity: its speakers detach and can be placed anywhere in a room.

The early, 1966 G2 has a T10 chassis made at Clairtone's factory in Stellarton, Nova Scotia, and a Garrard Lab 80 turntable.

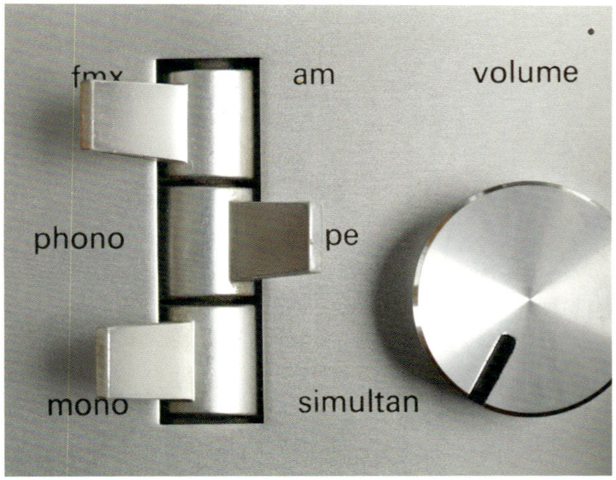

The later G2, released in 1968, has a T14 chassis, also made in Stellarton, and a Dual 1000 series or Garrard 75 turntable. As seen on pages 110 and 111, the turntable on the later G2 is set in the centre of the cabinet.

RIGHT **A service bulletin for dealers shows how to dismantle the G2 "sound globes."**

FAR RIGHT **The G2's 13-inch speakers, made of cast aluminum, are smaller, but heavier, and less easily dented than the G's spun-aluminum speakers.**

CLAIRTONE | SERVICE BULLETIN

CLAIRTONE SOUND CORPORATION LIMITED
Toronto New York Chicago Los Angeles

Ref: SB-6-7A — December 21, 1965.

To: All Clairtone Dealers, Service Companies and Representatives

From: E. A. Walker

Subject: Project G2 Sound Globes

Unlike the sound globes on the original Project "G", the sound globes used on the G2 can be dismantled and speakers exchanged in the field if it should be necessary.

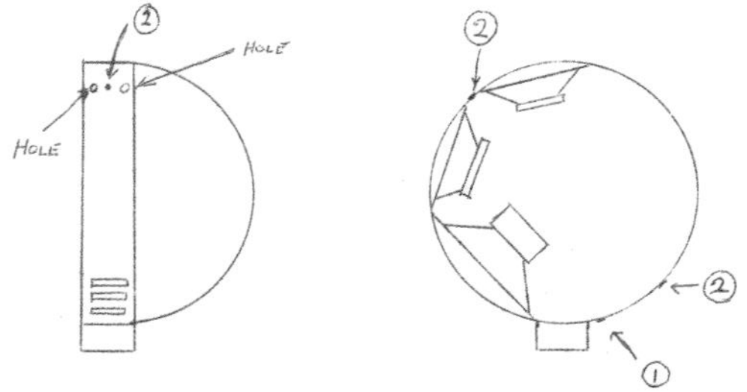

To dismantle the globe, remove screw (1). This allows the grille strip to be pulled back. After pulling back the grille strip, access to the Allen head tension screws (2) is possible. This should be slackened a few turns with a #5 Allen wrench to release tension on the internal spring clips. The Allen wrench or any other thin rod can then be pushed through one of the holes that are located on each side of the screw. Sufficient pressure should be used to lift the internal spring clip and then one half sphere can be removed.

This procedure is reversed to reassemble the globes but care should be exercised to replace all packing and gaskets as in original assembly.

E. A. Walker

EAW:ly

Head Office 100 Ronson Drive Rexdale Ontario telephone 249-1281 | New York Office 681 5th Avenue New York 22 EL 5-6600

LEFT Sean Connery with a G2, on the movie set of A Fine Madness (Warner Brothers, 1966).

RIGHT Jazz musician Bobby Hackett.

BELOW Singer and actor Dean Martin.

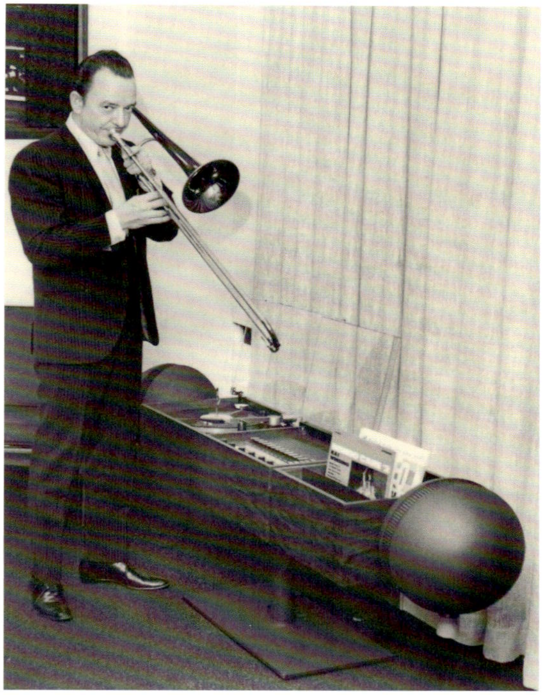

In this classic scene from The Graduate (United Artists, 1967), the G2 becomes an integral part of the seduction, as Mrs. Robinson (Anne Bancroft) mixes a drink for Benjamin (Dustin Hoffman) and puts on mood music.

Appealing to a conservative audience rather than the disco crowd, Go-Go dancer Katyha poses as an elegant chatelaine in these 1966 promotional images for the Contessa, Clairtone's faux Spanish cabinet introduced in 1963.

Clearly CLAIRTONE

Clairtone stereo and Clairtone color TV are designed to go together

7100 G-TV

7000 Project G

7002 G2

7001 G3

7101 Princess TV console

731 Princess stowaway

711 Princess

721 Signet

741 Signature

7102 Duchess TV console

743 Duchess

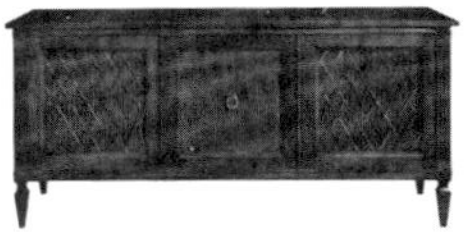
713 Palazzo

712 Patrician

732 Apollo stowaway

7104 Baroness TV console

714 Marquis

734 Baroness stowaway

7105 Contessa TV console

735 Infanta stowaway

745 Contessa

of Clairtone

"When you think of the company in holistic terms, you may find a clue to its eventual failure. Because the Clairtone that I know and love—Project G—represents such a small fraction of the business, there was a built-in contradiction. While Clairtone was marketing Project G, they were still shipping the clunky wood cabinets that define the Canadian vernacular 'stereo' of that period—in much greater quantity! **In some ways, Clairtone demonstrates our classically Canadian ambivalence: an unwillingness to fully embrace the new possibilities of design."**

Bruce Mau

DESIGNER, FOUNDER, BRUCE MAU DESIGN

A Clairtone brochure shows the entire 1967 product line.

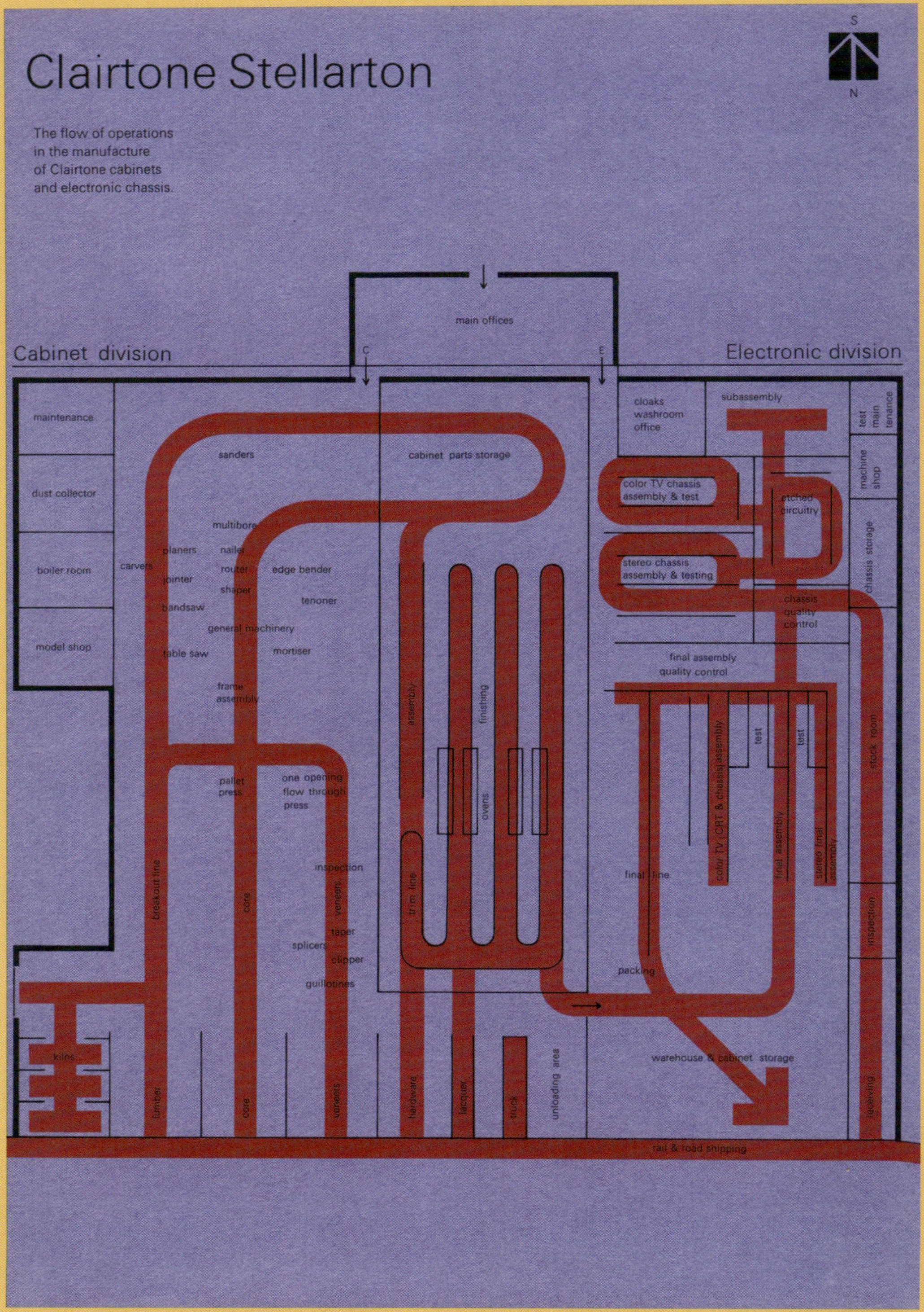

LEFT A schematic of the Stellarton factory.

RIGHT Nova Scotia premier Robert Stanfield (left) views a model of Clairtone's new plant, with Frank Sobey (centre), president of Industrial Estates Limited, and Peter Munk (right).

ERA THREE

On June 24, 1966, in the presence of 1,000 distinguished guests, Clairtone opened its state-of-the-art factory in Stellarton, Nova Scotia. Initially, all went well. For the first time in its history, Clairtone was making its own electronic components, instead of relying on subcontractors. As well, the G stereo line was expanded to include the G3, a tabletop model that reflected the trend towards modular hi-fi systems. Clairtone's new staff designer, Burton Kramer, introduced lowercase typography, and a comprehensive graphic design program.

The move to Stellarton also marked Clairtone's entry into the emerging field of colour television. Tony Mann, of Design Collaborative, created Clairtone's sculptural G-TV to complement the G line of stereos. Eight traditional TVs in wood cabinets were offered as well. But Clairtone, like others in the industry, had misjudged demand for colour TV, rushing into the market prematurely. At the Stellarton factory, thousands of unsold television sets soon piled up.

There were other problems at Stellarton, including labour unrest, a shortage of materials, and frequent stoppages on the assembly line. By the fall of 1967, Clairtone was on the edge of bankruptcy. Industrial Estates Limited, the provincial Crown corporation that financed the factory, agreed to bail out the company, but only on the condition that Peter Munk and David Gilmour relinquish control.

Munk and Gilmour were out. In their place was William Mangels, an American executive from Winchester Ammunition, and a former McKinsey consultant, whose push for cheap products and insensitivity to good design earned him a nickname among Clairtone employees: "the mangler." In late 1968, Mangels was replaced by George MacDonald, a Halifax businessman—to no avail. The Stellarton factory was shut down in 1971. Clairtone's assets were liquidated. All told, the Nova Scotia government lost around $25 million on Clairtone.

Clairtone's business story may be a tragedy, but its design legacy is anything but that. The company's passion for design—its dedication to contemporary consumer electronics—was unparalleled in North America 50 years ago. Clairtone's graphic design program, its marketing campaigns, and, of course, its unforgettable G line, look as modern today as they did when they were first released.

CN
CLAIRTON

Clairtone's huge, highly automated factory covered seven acres in the backwater of Stellarton, Nova Scotia. One third of the factory, or 100,000 square feet, housed the electronics division. The remaining space—200,000 square feet—was devoted to cabinetmaking. "[It] is not only big," remarked a Clairtone press release of 1966, "it's also the most modern, flexible and productive plant that has ever been built for this purpose anywhere." The building was designed by the Toronto architectural firm Crang & Boake.

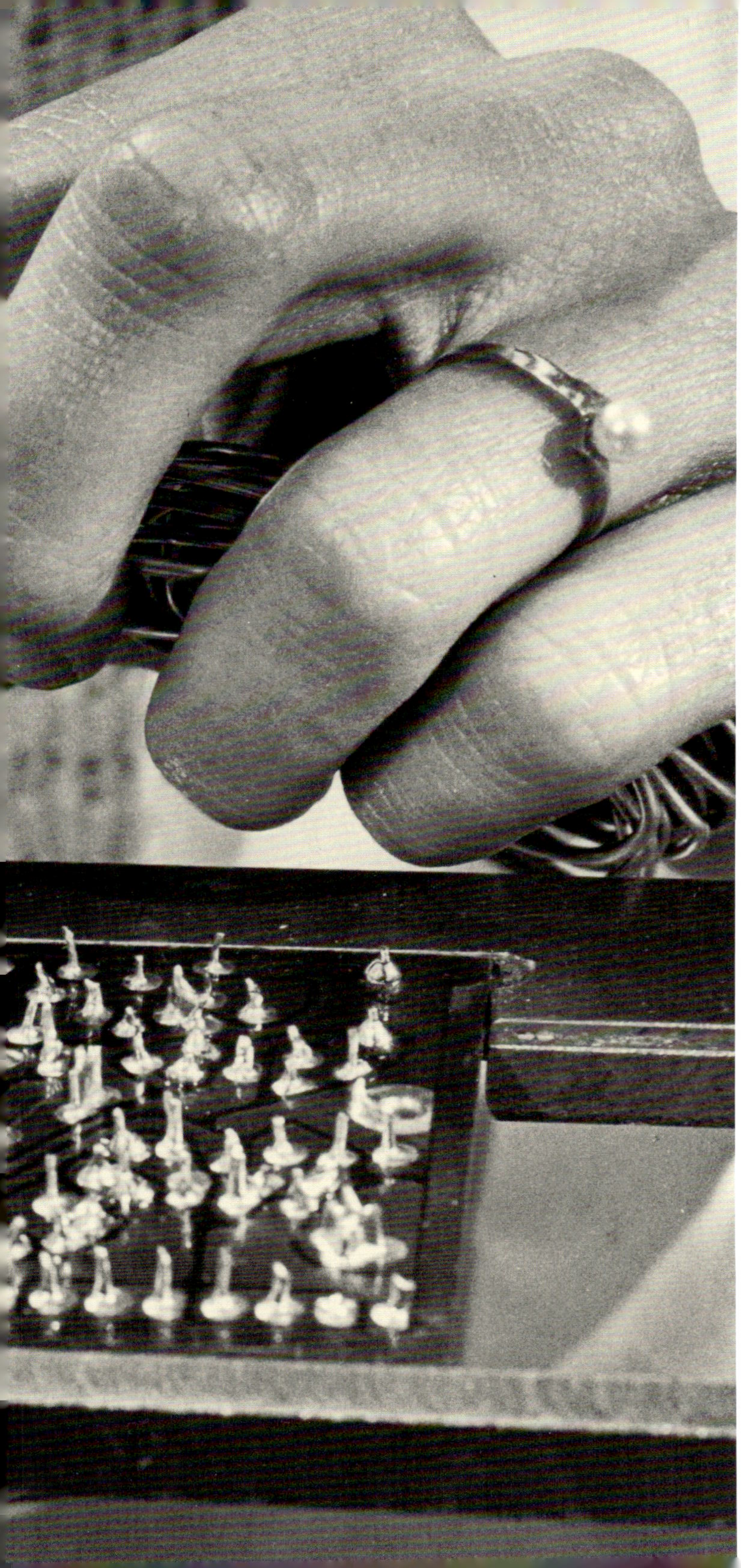

Women on the assembly line in Stellarton. "Clairtone's electronic testing and color convergence are completely handled by women," boasted a corporate press release. "The company's research has shown that women tend to be more dexterous, get bored less easily with repetitive jobs, and are more apt to maintain absolutely rigid inspection standards over long periods."

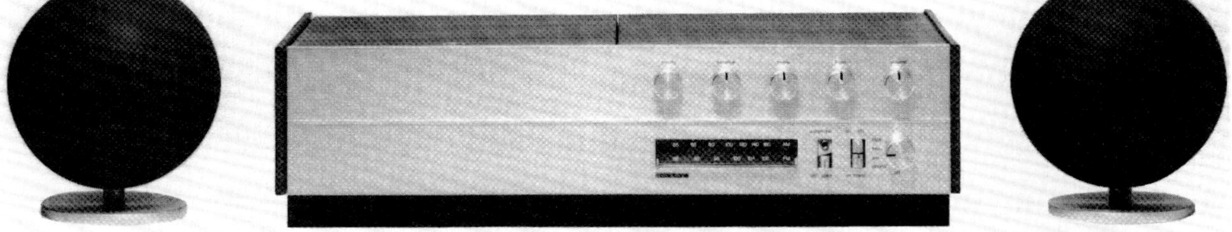

8250
Mini-G3

Mini-G is the most compact model in the now famous G Series. The modular design combines maximum versatility and performance in a space-conscious package. Consists of handsome walnut and brushed aluminum cabinet with two 9-1/2" Sound Globes mounted on circular aluminum bases. Sound Globes are finished in matte black.

Chassis: Clairtone All-Transistor T13C
Changer: Garrard 3000
Speakers: Sound Globes (9-1/2")
Finish: Walnut/Matte Black/Natural Aluminum
Style: Avant Garde
Length: (Cabinet) 30"
Depth: 14-3/4"
Height: 8-1/4"

G3

The smallest and least expensive of the G series, the G3 was created in-house, in consultation with Toronto's Design Collaborative. First sold in late 1966 with a matching bar cart for $499, the G3 was promoted in brochures as "mobile sound for young moderns." Its cabinet, made of brushed aluminum and birch veneer with a walnut finish, houses Clairtone's T16 chassis and a Garrard 3000 turntable. The "sound globes" are cast in plastic, and rest on elevated chrome rings. The Mini-G3, a variation of the G3, was displayed in Moshe Safdie's Habitat 67 at Montreal's Expo '67. It includes two "sound globes" mounted on aluminum pedestal bases, and a T13 chassis.

clairtone
tuning
balance
treble
bass
volume
spks off
afc on
fmx on
select on

"The Project G series did what Bang & Olufsen was about to do, i.e., it expressed the separation of elements utilizing palisander veneer and matt black paint. As a contemporary of Dieter Rams' seminal work for Braun, it remains a classic and should be elevated in the public consciousness once again."

Carl Magnusson

INDUSTRIAL DESIGNER, FORMER DESIGN DIRECTOR OF KNOLL INTERNATIONAL, FOUNDER, CARL GUSTAV MAGNUSSON DESIGN

Katyha and the G3.

Color TV–naturally–by CLAIRTONE
CLAIRTONE
remote control

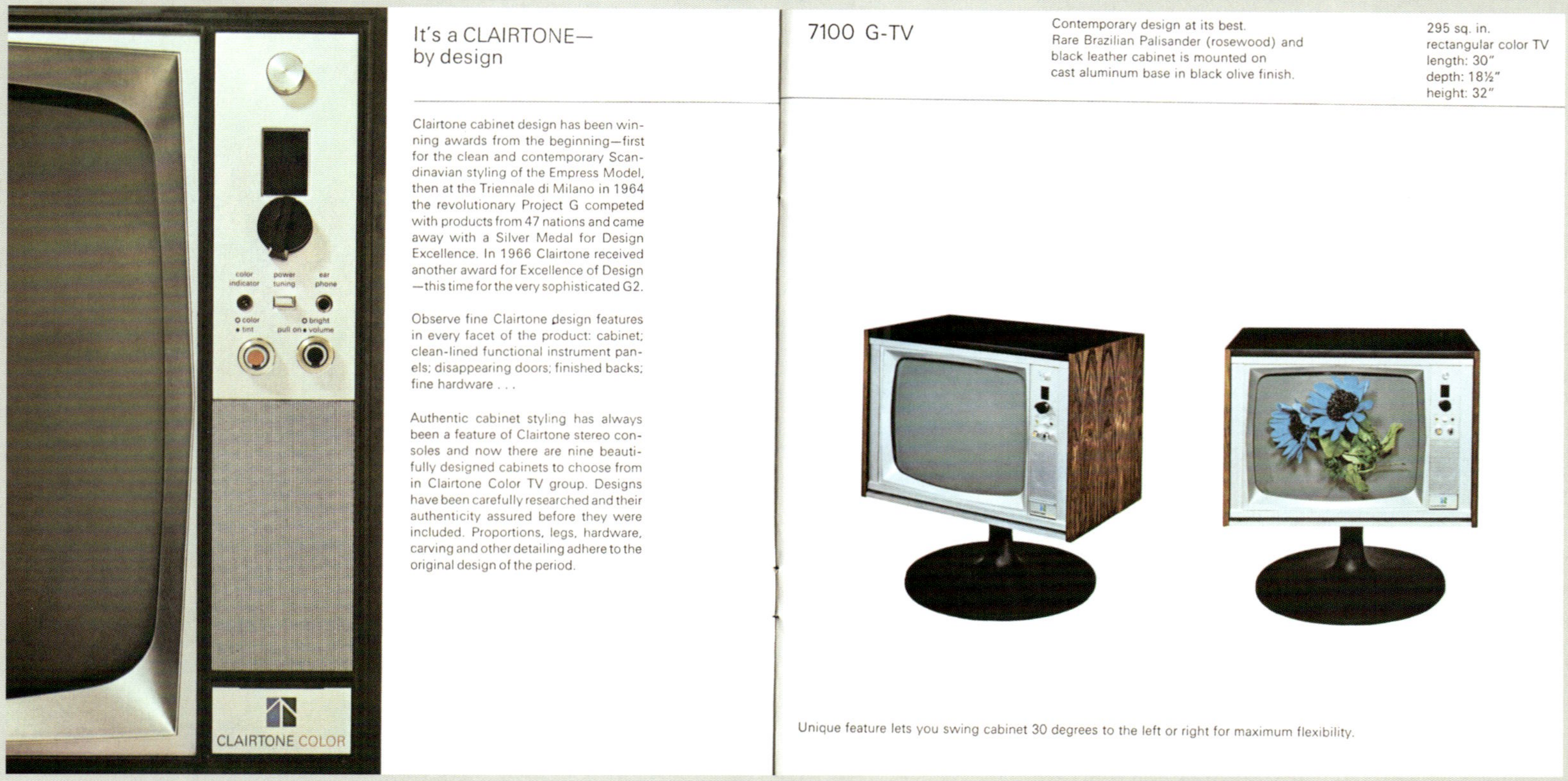

It's a CLAIRTONE—
by design

Clairtone cabinet design has been winning awards from the beginning—first for the clean and contemporary Scandinavian styling of the Empress Model, then at the Triennale di Milano in 1964 the revolutionary Project G competed with products from 47 nations and came away with a Silver Medal for Design Excellence. In 1966 Clairtone received another award for Excellence of Design—this time for the very sophisticated G2.

Observe fine Clairtone design features in every facet of the product: cabinet; clean-lined functional instrument panels; disappearing doors; finished backs; fine hardware . . .

Authentic cabinet styling has always been a feature of Clairtone stereo consoles and now there are nine beautifully designed cabinets to choose from in Clairtone Color TV group. Designs have been carefully researched and their authenticity assured before they were included. Proportions, legs, hardware, carving and other detailing adhere to the original design of the period.

7100 G-TV

Contemporary design at its best.
Rare Brazilian Palisander (rosewood) and black leather cabinet is mounted on cast aluminum base in black olive finish.

295 sq. in.
rectangular color TV
length: 30"
depth: 18½"
height: 32"

Unique feature lets you swing cabinet 30 degrees to the left or right for maximum flexibility.

Clairtone's 1966 colour TV brochure, designed by Frank Davies.

G-TV

Evoking Eero Saarinen's Tulip chair, the G-TV, designed by Tony Mann in 1966, rests on a modern pedestal base made of cast aluminum. The TV, composed of Clairtone's C15 chassis and a deluxe, 25-inch RCA picture tube, swivels, and is cased in rosewood and black leatherette to complement the G stereo series. In its day, the G-TV sold for $1,150, double the price of a basic colour TV from RCA Victor.

In an interview, Mann recalled mocking up the G-TV in a single weekend: "Peter [Munk] called me Friday at four in the afternoon from his car phone. It was my first experience with a car phone… There was a press conference on Tuesday and he wanted to launch the new television in a cabinet that didn't have a traditional look. Could I have something ready for Tuesday? he asked. I took an old bicycle wheel as reinforcement for the circular base, which I turned in plaster. It was unheard of to have such a short time frame, even today … but that was Peter."

START SPREADING THE NEWS

The "general theme" of this 60-second commercial, filmed by Fritz Spiess and written by Doug Linton for Young & Rubicam in 1967, was "elegance," according to the production notes. "Opening with car—elegant Rolls-Royce type in front, coming over bridge into the city as the sun comes up," the notes specify. "Traveling through deserted streets to an elegant store about to be opened, perhaps with men pulling down store canopy, rolling out carpet. Car pulls up at store, TV set is taken out, pull back as city comes alive, action, busy streets, people walking by." The bridge is the Brooklyn Bridge. The elegant store is B. Altman's, then on Manhattan's lower Fifth Avenue. As for the car, it's a restored 1936 Pierce-Arrow convertible with Peter Munk and David Gilmour up front and a G-TV in the back seat. This commercial, which won a gold medal at the New York International Film and TV Festival, was one of only two television commercials made by Clairtone.

RBER SHOP
the Empire

The Chatelaine model house at Expo '67 was "the perfect home for modern living," according to the official guide book to the World's Fair. Designed by the architect Gustavo da Roza, it resembled an oversized box made of wooden slats, its overhanging roof encasing the two-storey living-room area. The interior, with its unstructured layout, reflected the popular open-plan design of the 1960s. Notice Clairtone's G-TV in the living room (right). The Chatelaine exhibit, whose theme was "Man in His Home," was built entirely of Canadian materials, furnishings, and know-how.

With Katyha dressed to complement the television's leatherette side panels, this Clairtone advertisement in Chatelaine magazine confirms that the G-TV was chosen for the "Man in His Home" exhibit.

"THE FINEST COLOR TV IS A USED CLAIRTONE"

In the second, and final, Clairtone television commercial, Peter Munk and David Gilmour, seated in chairs by the Canadian designer Stefan Siwinski, attest to their company's high quality control. "The finest color TV you can buy is a used Clairtone," intones the voiceover, as engineers perform important technical work. "Only these two men ever see a new one... They will turn it on and use it for an equivalent viewing time of one week… Use it to ensure that every detail meets their standards of quality... Standards that revolutionized the look and sound of stereo… Standards uncompromising in building you the finest color TV. " This commercial of 1967, written by Doug Linton for Young & Rubicam, was filmed by Fritz Spiess at Toronto's TDF Artists studio and at Clairtone's Stellarton factory.

TONY MANN

ANTHONY "TONY" MANN (B. 1927)
Mann studied industrial design at the London Central School of Arts and Crafts. He spent a decade working as a graphic and exhibition designer in London before moving to Toronto in 1962, inspired, in part, by his friend Hugh Spencer. After joining the CBC, he was hired by Cooper & Beatty Ltd., where he replaced the distinguished typographer Allan Fleming as creative director. In 1965, with Al Faux, Mann founded the influential Design Collaborative in Toronto. Mann and Faux worked together on Clairtone's G2 and G-TV. The G-TV's well-proportioned panel—its small knobs reminiscent of a cockpit instrument panel—was influenced by Mann's research at Macy's in New York, where he tested TVs and interviewed TV salesmen and customers. In the late 1960s, Mann left commercial practice and founded the visual communication program at the Nova Scotia College of Art and Design, where he taught until his retirement in 1998.

Munk and Gilmour photographed with the G-TV prototype.

A 1968 instruction manual shows Mann's TV panel. It also features Clairtone's new logo, designed by Burton Kramer.

BURTON KRAMER

BURTON KRAMER (B. 1932)
Kramer is one of Canada's leading graphic designers, best-known for creating CBC's "exploding pizza" logo of 1974. Born in the U.S., he studied at the Institute of Design in Chicago, at Yale University, and, as a Fulbright scholar, at the Royal College of Art in London. In 1965, Kramer immigrated to Canada and joined Paul Arthur Associates, where he contributed to the navigational pictographs used at Expo '67.

Introduced to Peter Munk by Tony Mann, Kramer joined Clairtone as its in-house graphic designer. Most significantly, he changed Clairtone's graphic identity: his new lowercase logo, curvilinear and sans-serif, anticipated the popular Pump typeface of the 1970s. Kramer left Clairtone in 1968 to open his own firm specializing in visual communication. Among his clients were the Canadian Imperial Bank of Commerce, Teknion, and the St. Lawrence Centre for the Arts. Kramer taught corporate graphic design at the Ontario College of Art and Design for 21 years.

Kramer's logo, seen here on a Clairtone service van.

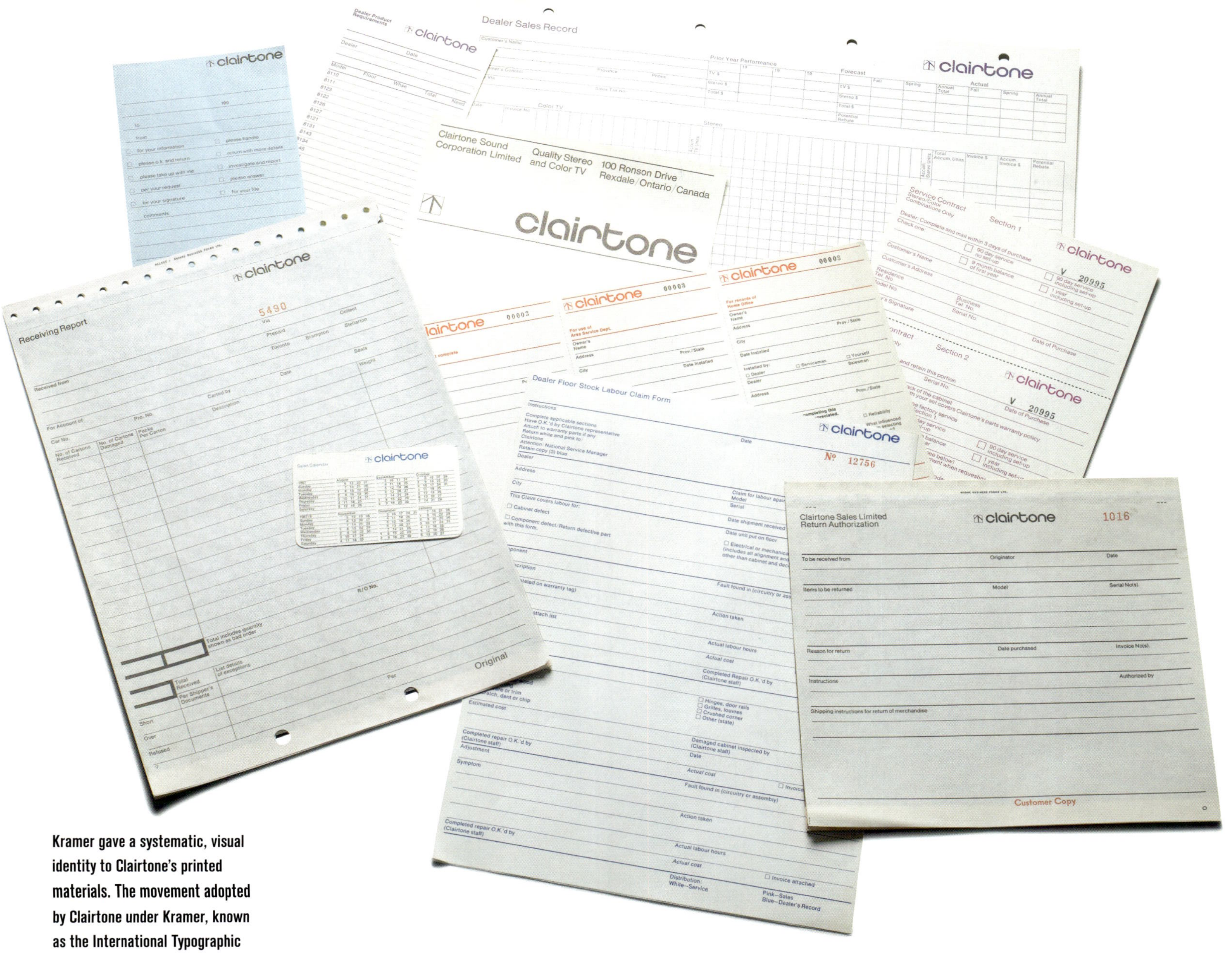

Kramer gave a systematic, visual identity to Clairtone's printed materials. The movement adopted by Clairtone under Kramer, known as the International Typographic Style, or Swiss Style, dictated that graphics and typography be functional, symmetrical, and organized on geometrical grids.

For Clairtone's 1966 annual report and price list, Kramer reinterpreted the company's arrow-shaped logo, experimenting with blue and purple geometric patterns that evoke Op Art.

Dealer Price List
August 1, 1967

Clairtone Sales
Limited

clairtone

SINCLAIR MICROMATIC
clairtone

MINI HI-FI

Introduced in 1968, the Mini Hi-fi is an almost identical copy of the 1967 Micromatic by England's Sinclair Radionics. Billed by both companies as "the world's smallest radio," the Mini Hi-fi and the Micromatic are the size of a matchbook and weigh one ounce (including batteries). For Clairtone, the Mini Hi-fi marked the end of an era: by the late 1960s, struggling to avoid bankruptcy, and under new management, the company pushed into the mass market with a strategy of low prices and high volume.

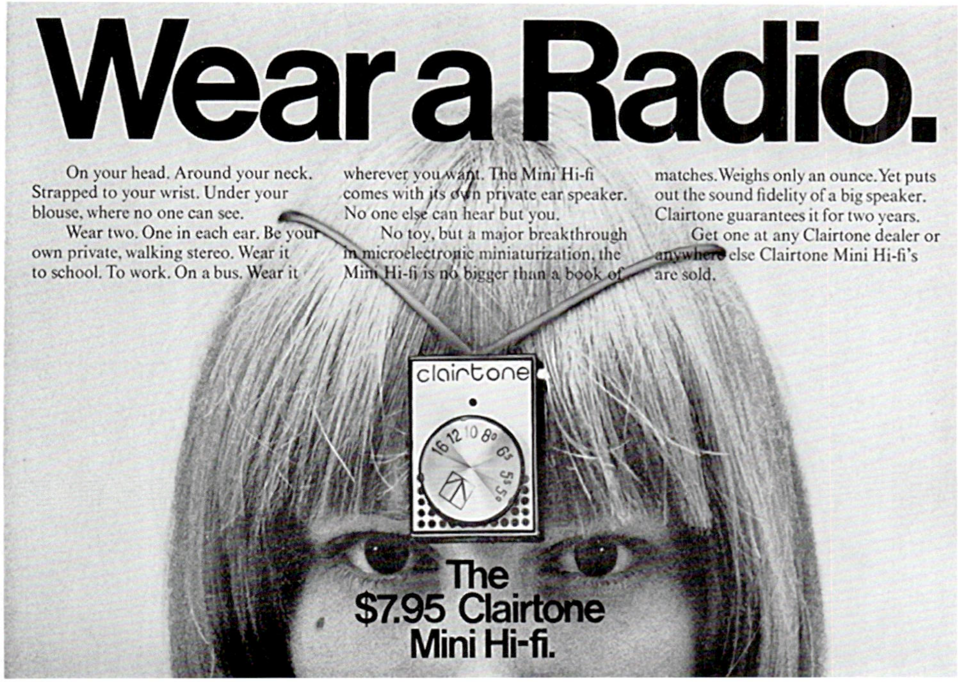

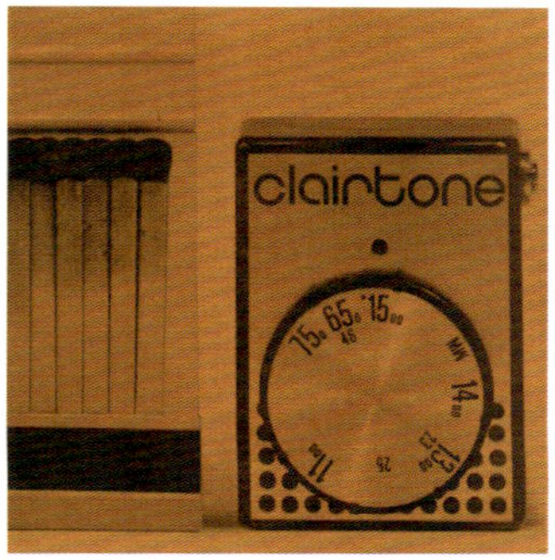

FAR LEFT **An advertisement for the Mini Hi-fi, by Lawrence Wolf (Canada) Inc. Rod Nash was the art director.**

ABOVE AND LEFT **Part of a Mini Hi-fi brochure.**

SIGHT OF SOUND

The Light Globe and Visual Translator was part of Clairtone's last original product line—the Sight of Sound collection, designed by Frank Davies. Although shabbily constructed, the collection is, in hindsight, a wonderful example of the Pop era. Connected to a stereo, the $400 Light Globe pulsated with the music, casting patterns of moving light on walls and ceilings, and creating an "experimental environment," in the words of Clairtone's P.R. department. Introduced at the 1968 Consumer Electronics Show in New York City, fewer than 500 Light Globes were ever made. Also shown here, on a tripod, is Clairtone's Light Projector Unit.

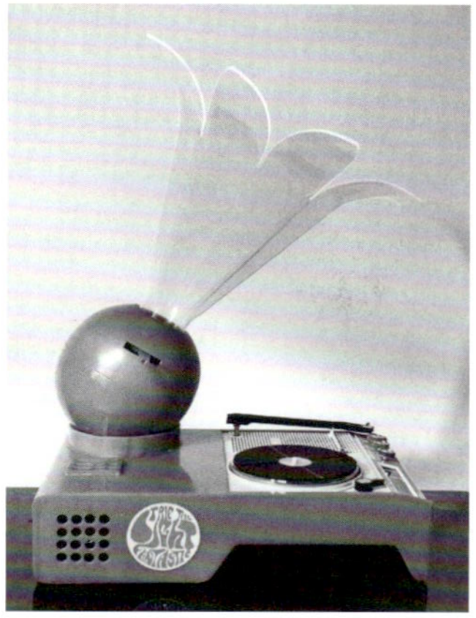

Described as a "psychedelic home kit," the $99 Light Fantastic was another piece of Frank Davies's collection of 1968. The two-speed portable phonograph and radio came with a transparent, Victrola-like plastic horn, which projected moving light patterns as the music played. Though it was cancelled almost as soon as it went into production, some people believed that Light Fantastic was a sure thing: "There are 200,000 hip bachelor pads in this city and within a year every one of them is going to have Light Fantastic," raved a 1968 issue of the Star Weekly.

A model showroom on rails, the Clairtone Train travelled right across Canada, bringing the company's line of products to dealers in small towns. As a sales technique, the Clairtone Train was innovative. "It was about creating a 'total experience' for the buyers," said Eleanor Brydone, then a staff designer with A.D. Pollard & Assoc., the firm that designed the showroom. Still, it was a long way from the days when Clairtone flew department-store executives to Stellarton on the company plane. "We were just trying to hold things together," is how Jack Lane, a longtime Clairtone employee, remembered the train. "We were fighting a losing battle." These drawings were published in Canadian Interiors, January 1970.

2 3 4→

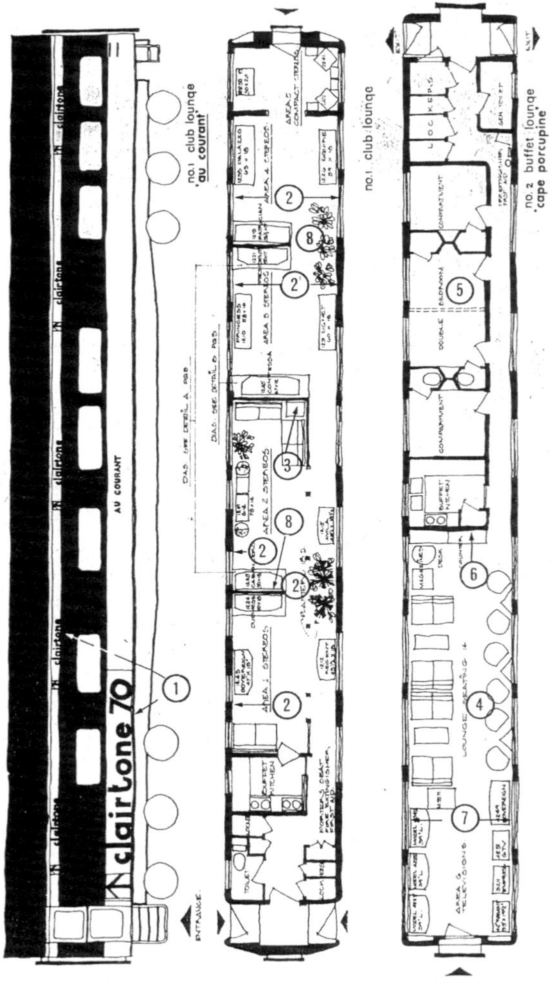

LEGEND FOR INTERIORS OF CLAIRTONE 70 TRAIN

1. *Train Exterior:*

Blue Canopy over entrance way.
Large groupings of flying balloons to be corporate colours.
Exterior graphics in Clairtone corporate purple.

2. *Train Interiors:*

Clairtone cabinet display areas.
Heavily textured natural wool draperies.
Plantings distributed for colour, texture and acoustical values.
Red carpeting.
Concealed floor mounted spot lights — to dramatize equipment.

3. *Clairtone Graphics:*

in Clairtone purple on white plastic panel on black ground. note 'Clairtone' to run off of panel as illustrated.

4. *Bar Lounge Area:*

'Logo' between window panels in purple.
Table accessories to have Clairtone 'Logo' imprinted in purple on white.
other bar accessories in purple.
Waiters to be in white jackets, purple trousers.
Seating upholstery — charcoal grey.
Red carpeting.
Textured draperies.

5. *Double Compartment:*

converted to office area.
White top desk with chrome base.
Desk chairs to be purple.
Red carpeting.
Repeat heavily textured draperies.
Logo repeated on panels between windows.

6. *Painting on Wall:*

Gerald Sevier original.

7. *Television Display Area:*

window walls covered with shag carpet, primary yellow.
Note—for sound and light control. Desire dramatic effect in this area before entering lounge area.

8. *Dais for Equipment Display:*

basic 'T' shape construction and covered in purple felt.
Sevier original on panel above.
Note—paintings selected have predominance of primary colours.

Design Group
A. D. POLLARD & ASSOCIATES LIMITED

Staff Designer
Eleanor Brydone A.O.C.A.

DOLLAR SALES
58 59 60
62 63 64 65
64 65

"The leadership in Clairtone understood design. They understood that creativity was the key to innovation. They were decades ahead of their time."

J. Eleanor Brydone

INTERIOR DESIGNER, FOUNDING PRINCIPAL, RICE BRYDONE

Peter Munk and David Gilmour, 1966. A G3 speaker is on the desk.

ACKNOWLEDGEMENTS

We are profoundly indebted to the former employees and executives of Clairtone Sound Corporation. Without their help, this book would not have been possible. Anna Chester (formerly Anna Spencer), Michael Chojnacki, Gloria Collinson, David Gilmour, Jack Lane, Anthony "Tony" Mann, and George Weil not only agreed to be interviewed—they also answered our endless questions and went far out of their way to find vital documents and photographs. Others connected to Clairtone provided us with valuable information: Leslie Berenyi, Frank Davies, Arthur Dollery, Stan Freeman, Steven Friedlich, Malcolm "Mac" Hogarth, Vladimir "Willy" Jidkoff, Burton Kramer, Doug Linton, Robert Manuge, Cubby Marcus, Jerome Markson, Morley Markson, Anna McCowan Johnson (formerly Anna Gilmour), Ian McLeod, Sol Mandlsohn, Gary Oldfin, David Pols, and Greg Yaneff.

From the beginning, Peter Munk supported us without reservation, offering enthusiasm, candour, and unrestricted access to his personal files. His contribution was invaluable. Graciously, DC Hillier and Daniel Leblanc lent us many items from their extensive collection, shared their archival material, and helped identify Clairtone's electronic components.

At Design Exchange, Samantha Sannella was the leading advocate of this book. Daniela Bryson and Courtney Stewart spent countless hours managing the Clairtone archives, researching historical images and articles, tracking down sources, and diligently checking every credit. Curator Michael Prokopow contributed his expertise.

We're also indebted to Doug Pepper and Jenny Bradshaw at McClelland & Stewart; Garry Shutlak of the Nova Scotia Archives and Records Management; and the ever-resourceful Sheila Fennessy. George Whiteside worked under a tight deadline, and an even tighter budget, to produce the outstanding new photographs in this book. Tyler Brûlé, Shirley Blumberg, Eleanor Brydone, Douglas Coupland, Carl Magnusson, Bruce Mau, Glenn Pushelberg, and Karim Rashid generously agreed to play cameo roles by putting their thoughts and ideas on paper.

Linda Munk, Professor Emeritus at the University of Toronto, was our fastidious editor. She simplified our prose and corrected more typos than we care to remember. Bruce Handy of *Vanity Fair* read an early version of the manuscript. His thoughtful suggestions made this a better book.

Our deepest gratitude goes to Barb Woolley and her extraordinary design team at Hambly & Woolley, including Emese Ungar-Walker, Sonja Kloss, and Ben Rivers. This book is as much theirs as it is ours.

NOTES AND SOURCES

For the most part, research for this book derives from dozens of interviews conducted mainly in 2006 and 2007. Our research also depended on the rich collection of inter-office memos, confidential reports, blueprints, ledgers, minutes, letters, brochures, annual reports, newsletters, and other historic Clairtone documents stored at the Nova Scotia Archives and Records Management (NSARM) in Halifax and at Design Exchange (DX) in Toronto. Two books were key sources: Garth Hopkins's *Clairtone: The Rise and Fall of a Business Empire* (McClelland & Stewart, 1978), and Harry Bruce's *Frank Sobey: The Man and the Empire* (Macmillan of Canada, 1985).

Peter Munk, David Gilmour, and the Memory of Clairtone

19 Clairtone was his "first love": Dean Walker, "Peter Munk Plots a Comeback," *Executive*, December 1978.

19 "Clairtone was the single most formative experience": Clyde Farnsworth, "Peter Munk: Lucky Gold Strike and Wise Hedging Keep His Shareholders Smiling," *New York Times*, May 30, 1993.

20 "Everybody knew about Clairtone": Joan Sutton, "Peter Munk: Success to Failure to Success," *Toronto Star*, January 19, 1981.

20 "the TSE was as WASP a club as you can get": Telephone interview by Nina Munk, July 26, 2007.

20 At Simpson's, one Clairtone hi-fi was sold every three hours: *Clairtone News*, March 1966. NSARM, Clairtone fonds MG 3, vol. 2164, file 11.

20 "I worked 70- and 80-hour weeks": Telephone interview by Nina Munk, July 26, 2007.

21 "Canadians would have popped a few buttons": "New Canadian Firm Making Mark at Home and in U.S.," *Globe and Mail*, April 9, 1959.

21 "everybody's darlings": Alexander Ross, "How Peter Munk Was Sunk by a Euphoria of Success," *Financial Post*, November 1, 1969.

21 "They were treated as movie magazines treated Rock Hudson": Harry Bruce, *Frank Sobey: The Man and the Empire*, p. 289.

21 "There was a year when I had 34 speaking engagements": *the fifth estate*, CBC News, 1978.

21 "The idea was good": *Toronto File*, CBC TV, January 25, 1965.

24 "As men carry set through to entrance of store": "Production Notes 'New York' Clairtone Commercial," photocopy of notes, presumed to be from Young & Rubicam, July 7, 1967. DX Clairtone archives.

24 "Over the course of two or three meetings with Peter and David, I became so totally imbued": From a 1975 interview quoted in Garth Hopkins, *Clairtone: The Rise and Fall of a Business Empire*, pp. 42-43.

26 Bruce Mau: In an email to Nina Munk, August 23, 2007, Mau wrote: "In the end, Clairtone failed to go to the future with commitment. Project G was a marketing project, instead of a philosophy."

26 No more than a few hundred Project Gs were ever made: In one of the few references to the number of Project Gs manufactured, a February 5, 1964 memo from Peter Munk to Zigmund Hahn, Clairtone's chief financial officer, states that the "initial budget allocation to 'G' for publicity, promotions and advertising was made on a basis of producing 150 units, but now this figure … has been increased to 400." NSARM, Clairtone fonds MG 3, vol. 2151, file 10. Another reference, a January 27, 1965 memo from Michael Chojnacki to George Weil, states: "It appears that we shall have approximately 360 [Project G] cabinets in stock by the end of July." NSARM, Clairtone fonds MG 3, vol. 2172, file 5.

26 "Trend setters" were the target market: "Project G," undated internal report from Clairtone Electronic Corporation (the U.S. subsidiary). The report's contents make clear it was written in late 1963. NSARM, Clairtone fonds MG 3, vol. 2149, file 4.

27 "You may have to stage a 'publicity stunt'": "Publicity Procedure for Project G Openings," inter-office memo from David Gilmour to all U.S. regional managers and salesmen, February 11, 1964. NSARM, Clairtone fonds MG 3, vol. 2209, file 1.

27 "design something 'way out'": Wynne Thomas, "It Sells in the U.S. with a 'Touch of Mars,'" *Marketing*, January 31, 1964.

27 "I remember the day": Joint interview by Nina Munk, New York City, September 17, 2006.

28 "I thought teak looked too plain": Telephone interview by Nina Munk, August 3, 2007.

28 Oscar Peterson affirmed publicly that his music sounded as good on a Project G as it did live: "Designer Notebook: By Accident and Design," *Canadian Interiors*, May 1964.

29 "It took the patience of Job": Interview by Nina Munk, New York City, September 18, 2006.

29 "Oh, we had some fights": Telephone interview by Nina Munk, August 3, 2007.

29 "Hugh was temperamental": Interview by Nina Munk and Rachel Gotlieb, Toronto, May 18, 2007.

29 "I needed Hugh Spencer": Ibid.

29 Inter-office memos question the effectiveness of Clairtone's advertising: In "U.S. Sales," an October 18, 1963 memo to David Gilmour, Peter Munk notes that a prominent ad campaign in *Life* resulted in just nine units sold. NSARM, Clairtone fonds MG 3, vol. 2151, file 9. In an untitled November 7, 1963 memo to Gilmour, Zigmund Hahn, Clairtone's chief financial officer, writes: "Judging from sales, the [advertising] program has been a failure." NSARM, Clairtone fonds MG 3, vol. 2149, file 4. In "Merchandising in USA 1965," a November 19, 1963 memo to Gilmour, Leslie Berenyi, a Clairtone VP, complains that sales in U.S. department stores are not worth the cost of co-op ads: "We were making one concession after the other yet business is gradually coming to a stand still." NSARM, Clairtone fonds, MG 3, vol. 2149, file 4.

31 Clairtone threatens to sue General Instrument of Canada: Beginning in the summer of 1963, Clairtone and General Instrument exchange a series of increasingly heated letters authored by Peter Munk, Malcolm Hogarth, Clairtone's director of engineering, and Fred Shuh, General Instrument of Canada's general manager. Finally, in a letter dated November 28, 1963, Hogarth threatens legal action: "The purpose of this letter is to request a reply from you … prior to instructing our lawyers to start proceedings." NSARM, Clairtone fonds, MG 3, vol. 2150, file 5.

31 "Emergency measures" are needed: Untitled inter-office memo, February 6, 1963, from Zigmund Hahn to Peter Munk. NSARM, Clairtone fonds, MG 3, vol. 2149, file 12.

31 "It was an impossible situation": Interview by Nina Munk, September 3, 2007.

32 "I'll never forget that flight": Telephone interview by Nina Munk, August 3, 2007.

32 "This arrangement is absolutely fantastic!": "Clairtone Planning N.S. Branch to Produce Television Sets," *Globe and Mail*, November 19, 1964.

33 An automated dust-collection system: Alfred de Bat, "Is Clairtone the Most Modern Furniture Plant in the Western Hemisphere?" *Furniture Design & Manufacturing*, October 1966.

33 A miraculous seven-headed moulder: "The Cabinet Factory," undated Clairtone press release (#4 of 8 in a series "The Clairtone Story"). Collection of George Weil.

33 "It can take a piece of raw wood": Alfred de Bat, "Is Clairtone the Most Modern Furniture Plant in the Western Hemisphere?" *Furniture Design & Manufacturing*, October 1966.

33 Intense bulbs "increase productivity by eight percent": "The Electronic Plant," undated Clairtone press release (#3 of 8 in a series "The Clairtone Story"). Collection of George Weil.

33 "Munk was too good a salesman for his own good": Harry Bruce, *Frank Sobey: The Man and the Empire*, p. 291.

33 "Stellarton's prospects were bleak": Ibid, pp. 243-44.

34 "The general population is basically not geared to the manufacturing frenzy": "Report on the Stellarton Operation," Clairtone internal report, September 5, 1967. Other internal reports detailing Clairtone's problems in Nova Scotia include "Financial Comments—1967," September 28, 1967 and "Clairtone Stellarton Plant," October 31, 1967. (None of the authors is identified.) NSARM, Clairtone fonds MG 3, vol. 2159, file 3.

34 "It is unbelievable and very disturbing": "Material Shortages," inter-office memo from Michael Chojnacki to J. Croft, July 19, 1967. NSARM, Clairtone fonds MG 3, vol. 2156, file 16.

34 "We are still unable to commence delivery of TV sets": "Production Delays—1968 models," inter-office memo from Michael Chojnacki to R.C. "Ray" Jones, July 19, 1967. Ibid.

34 "Chaos": Peter Munk, testifying before the Standing Committee of the Nova Scotia Legislature on Industry, House of the Assembly, Halifax, March 18, 1969; quoted in "'Help That Never Came' was Blow to Clairtone," *Toronto Daily Star*, March 19, 1969.

35 "Success Story Turns Sour": "The Clairtone Saga: A Canadian Success Story Turns Sour," *Toronto Daily Star*, October 3, 1967.

37 "stereophonic earphones": Clairtone brochure, 1963. DX, Clairtone archive.

37 "I ended up with less money than the mortgage was worth": Dean Walker, "Peter Munk Plots a Comeback," *Executive*, December 1978.

37 "It's a good thing we have people like Peter Munk in Canada": Frank Sobey testifying before the Standing Committee of the Nova Scotia Legislature on Industry, House of the Assembly, Halifax, March 22, 1968 quoted in Garth Hopkins, *Clairtone: The Rise and Fall of a Business Empire*, pp. 204-05.

Era One

38 Simpson's was Clairtone's biggest retailer: See source for p. 20.

39 "We went into Italian provincial and French provincial": Interview by Nina Munk, New York City, September 18, 2006.

41 "I wore a dress that I saw in *Vogue*": Telephone interview by Rachel Gotlieb, August 13, 2007.

42 Dalton Camp campaign: Richard Flohil, "Simple, Believable Approach Said Best," *Globe and Mail*, May 4, 1962; and Tim Dickson, "Posh Style from Music is Paying Off," *Marketing*, March 31, 1961.

42 "Peter and David were [initially] unhappy": Telephone interview by Rachel Gotlieb, May 22, 2007.

46 "The pattern originally derives from the photographic track": Frank Davies, notes written for DX exhibition "Pop in Orbit," 1995. DX, Frank Davies file.

46 "One of the first two-storey trade booths in Canada": Gordon McCaffrey, "A Booth at the Ex—Is it Worth it?" *Toronto Daily Star*, August 22, 1962; and Anna Chester, profile of Hugh Spencer written for the DX exhibition "Pop in Orbit," 1995. DX, Hugh Spencer archive.

48 "ultra-modern": Clairtone annual report, 1962. DX Clairtone archive.

55 "Try appealing to women": David Fulton, "How a small company takes on the giants," *Executive*, August 1961.

59 "few people had heard of Braun": Telephone interview by Rachel Gotlieb, August 3, 2007.

59 Braun cancels contract: A January 16, 1964 registered letter from Braun executives (signatures illegible) states: "Your purchase figures for 1962 and 1963 are below minimum…Under these circumstances we are not able to continue the contract of January 3, 1962." NSARM, Clairtone fonds MG 3, vol. 2219, file 13.

60 "A visual method of communicating an ideal": Clairtone annual report, 1961, DX Clairtone archive.

65 Project G "provided a true sound cone projection": "Project G Fact Sheet," undated press release, collection of Michael Chojnacki.

66 "What is so remarkable about Project G": Telephone interview by Rachel Gotlieb, August 23, 2007.

66 For records on Project G, see: NSARM, Clairtone fonds MG 3, vol. 2172, file 4, and vol. 2209, file 1.

66 "When we approached Supreme": Wynne Thomas, "It Sells in the U.S. with a 'Touch of Mars,'" *Marketing*, January 31, 1964.

71 Rosewood had a "plastic appearance": "Project 'G' Cabinets," inter-office memo from David Gilmour to Michael Chojnacki, March 29, 1965. NSARM, Clairtone fonds MG 3, vol. 2172, file 5.

72 "They were inseparable": Telephone interview by Rachel Gotlieb, June 14, 2007.

72 "For Hugh, purple was dark grey": Telephone interview by Rachel Gotlieb, August 3, 2007.

72 "He believed purple was a very architectural colour": Email to Rachel Gotlieb, November 22, 2007.

79 "Canada means wheat, fish, and endless silent forests": "Canadian Hi-Fi Maker Cracks US Market," *Business Week*, March 24, 1962.

79 "Mr. Morton May of May Department Stores": Interview by Nina Munk, New York City, September 18, 2006.

79 Only a single working prototype: "Project 'G'," inter-office memo from Malcolm Hogarth to S. Stephenson, November 12, 1963, states: "The new project 'G' is being forwarded to the New York Office.… It is the one and only prototype of its kind." NSARM, Clairtone fonds MG 3, vol. 2150, file 5.

80 "We took the prototype and strapped it on": Email to Rachel Gotlieb, July 30, 2007.

83 "To me it was one of the greatest designs": Interview by Nina Munk, New York City, September 18, 2006.

87 3,000,000 people passed through the show: *Clairtone News*, September 1965. DX Clairtone archives.

93 Sinatra gave away half a dozen: Michael Barkway, "A look at Canada's Bid to Improve Its World Position," *Canada: Operation World Markets*, September 1963; and "Taking on the Big Names in Hi-Fi," *Printer's Ink*, January 1964.

95 Canadian cabin in Milan: Madge Phillips, "Triennale Canadian Exhibit," *Canadian Interiors*, August 1964.

Era Two

96 Clairtone sponsored Miss Canada: Miss Canada Pageant booklet, 1966. Library and Archives Canada, June Dennis fonds MG31 D191, vol. 1. And *Clairtone News*, December 1964. DX Clairtone archive.

97 "There was always the intention for a G collection": Interview by Nina Munk and Rachel Gotlieb, Toronto, May 18, 2007.

97 "The G was the Thunderbird": Telephone interview by Rachel Gotlieb, August 3, 2007.

97 Clairtone's line included "everything from Spanish to Danish": "Absolutely Shocking" advertising campaign, 1965. DX Clairtone archive.

100 Macy's was one of Clairtone's most important U.S. retailers: *Clairtone News*, September 1965. DX Clairtone archive.

104 "a striking example of contemporary freedom of taste": Ray Faulkner and Sarah Faulkner, *Inside Today's Home*, Third Edition (Holt, Rinehart and Winston, 1968), p. 120.

111 For records on G2, see: NSARM, Clairtone fonds MG 3, vol. 2170, file 9, and vol. 228, file 13.

116 "Project G's speakers were too big": Telephone interview by Rachel Gotlieb, October 17, 2007.

122 "Eight G2 column-mounted Sound Globes and Katyha": *Clairtone News*, March 1966. DX Clairtone archive.

123 "Sorry we can't be with you—but have fun": Clairtone promotional letter, undated. Collection of Daniel Leblanc and DC Hillier.

Era Three

131 Problems at Stellarton: See sources for p. 34.

131 William Mangels: John Picton, "Man in the News: Challenge draws Clairtone's new Chief," *Globe and Mail*, February 28, 1968; and Garth Hopkins, *Clairtone: The Rise and Fall of a Business Empire*, pp. 131-32.

133 "[It] is not only big": "History of the New Plant," undated Clairtone press release (#2 of 8 in a series "The Clairtone Story"). Collection of George Weil.

135 "completely handled by women": "The Electronic Plant," undated Clairtone press release (#3 of 8 in a series "The Clairtone Story"). Collection of George Weil.

141 "Peter [Munk] called me Friday at four in the afternoon": Telephone interview by Rachel Gotlieb, August 3, 2007.

142 The "general theme" was "elegance": "Production Notes 'New York'," photocopy of notes, presumed to be from Young & Rubicam, July 7, 1967. DX Clairtone archives.

147 Everything "on screen" was "black-white": "Designer's Notebook: Clairtone Exhibit is Ingenious and Flexible," *Canadian Interiors*, January 1967.

148 *Chatelaine* model house: Alan Campaigne, "Chatelaine Expo Home," *Chatelaine*, January 1967.

151 "The finest color TV you can buy": "Production Notes 'Clairtone Factory'," photocopy of notes, presumed to be from Young & Rubicam, July 7, 1967. DX Clairtone archives.

159 For records on the Mini Hi-fi, see: NSARM, Clairtone fonds, MG 3, vol. 2182, files 17-18.

161 An "experimental environment": "The Sight of Sound," undated Clairtone press release, collection of Gloria Collinson and Frank Davies. For records on the Sight of Sound, see: NSARM, Clairtone fonds MG 3, vol. 2189, file 2.

163 "There are 200,000 hip bachelor pads": Barrie Hale, "The New $99 Home Psychedelic Kit," *Star Weekly*, August 10, 1968.

164 "It was about creating a 'total experience'": Telephone interview by Rachel Gotlieb, June 8, 2007.

164 "We were just trying to hold things together": Telephone Interview by Nina Munk, July 31, 2007.

CREDITS

Every effort has been made to locate and obtain permission from copyright owners. Please contact Design Exchange to report possible oversights. Unless otherwise noted, "DX" refers to Design Exchange, Toronto, Clairtone archive. And "NSARM" refers to the Nova Scotia Archives and Records Management, Halifax, Clairtone fonds MG 3. When available, file numbers or collection identification numbers are given in parentheses.

cover Collection of Daniel Leblanc and DC Hillier. Photo by George Whiteside/© DX.

i-ii Collection of Daniel Leblanc and DC Hillier. Photo by George Whiteside/© DX.

iii From Clairtone "Absolutely Shocking" ad, 1965. DX.

1 NSARM (vol. 2183, file 15).

2-3 Courtesy of Warner Bros.

4-5 From Clairtone annual report, 1963. DX.

6 From Opus International brochure. DX, Hugh Spencer archive.

8 Courtesy of George Weil.

9 From *Executive*, August 1961.

11 DX. Photo by Jack Marshall & Co. Ltd.

12-13 Collection of Daniel Leblanc and DC Hillier. Photo by George Whiteside/© DX.

14 From Clairtone annual report, 1962. DX.

16 DX. Photo by Everett Roseborough.

18 From "Clairtone Sound Corporation: A Background Report," W.E. Hutton & Co., 1966. DX.

20 DX. Photo by Graphic Artists.

21 (top) DX. Photo by Panda Photography. (bottom) Courtesy of George Weil.

22 DX.

23 (left) From "It Was Time the World Knew Them/Profitable PR Chris Yaneff," undated/unsourced reprint. Courtesy of Greg Yaneff. (right) From Clairtone annual report, 1960. DX.

24 Courtesy of Peter Munk.

25 NSARM (clockwise, from top left: vol. 2151, file 9; vol. 2149, file 6; vol. 2151, file 10).

26 NSARM (vol. 2225, file 14). Photo by Warren Bowen.

27 (left) Collection of Daniel Leblanc and DC Hillier. Photo by George Whiteside/© DX. (right) Courtesy of Peter Munk.

28 Courtesy of *Playboy*/© 1964 by *Playboy*.

29 Courtesy of the Walt Disney Corp.

30-31 DX.

32 (top) Courtesy of Steven Friedlich. (bottom/left) Courtesy of Gloria Collinson. Photo by George Hunter.

33 Courtesy of George Weil.

34 Courtesy of Peter Munk.

35 (left) Courtesy of Peter Munk. (right) Courtesy of Metro-Goldwyn-Mayer Inc.

36 Clairtone annual report, 1964. DX.

37 Courtesy of Nina Munk.

Era One

38 From *Canadian Homes and Gardens*, January 1960.

39 (top) From *Toronto Stock Exchange Review*, April 1963. DX. (bottom) *New York Times*, September 27, 1964.

40-41 From Clairtone brochure, 1958. DX.

42-43 Courtesy of Peter Munk.

44 From Clairtone annual report, 1960. DX.

45 DX (2007.14). Photo by George Whiteside/© DX.

46 (left) Robertson Davies Library, Massey College, University of Toronto. Photo by Legg Brothers Limited. (right) Collection of Daniel Leblanc and DC Hillier. Photo by George Whiteside/© DX.

47 DX.

48-49 Courtesy of George Weil.

50 (left) Courtesy of Greg Yaneff. (right) Courtesy of Greg Yaneff.

51 From Clairtone annual report, 1960. DX.

52 DX.

53 NSARM (vol. 2225, file 15). Photo by Skinner, Quin & Saabor Limited.

54-55 NSARM (vol. 2149, file 1). Photo by Everett Roseborough.

56 DX.

57 DX.

58 DX. Photo by Everett Roseborough.

59 (top) NSARM (vol. 2222, file 7). Photo by Everett Roseborough. (bottom left and bottom centre) Courtesy of Peter Munk. (bottom right) Collection of Michael Chojnacki. Photo by George Whiteside/© DX.

60 (top) DX, Hugh Spencer archive. (bottom) DX.

61 From Clairtone annual report, 1961. DX.

62-63 Collection of Daniel Leblanc and DC Hillier. Photo by George Whiteside/© DX.

64 Collection of Daniel Leblanc and DC Hillier. Photo by George Whiteside/© DX.

66-69 Collection of Daniel Leblanc and DC Hillier. Photos by George Whiteside/© DX.

70-71 DX (998.9). Gift of Peter Munk. Photos by George Whiteside/© DX.

72-73 Courtesy of *Canadian Interiors*.

74 Courtesy of George Weil.

75 From *Canadian Business*, February 1965.

76-77 Courtesy of Michael Chojnacki.

78 NSARM (vol. 2225, file 14). Photo by Herb Nott & Co. Ltd.

79 (top) NSARM (vol. 2225, file 14). Photo by Sonnenburg Fotografi. (bottom) NSARM (vol. 2225, file 14). Photo by Everett Roseborough.

80-81 DX.

82 From "Consider Project G" brochure. Courtesy of Michael Chojnacki. Photo by Everett Roseborough.

83 (top left) Clairtone brochure, 1963. DX. (bottom left) From *Financial Post*, November 1, 1969. (right) Courtesy of Nina Munk.

84 Courtesy of Gloria Collinson.

85 From *Clairtone News*, March 1965. DX.

86 Courtesy of Gloria Collinson.

88-89 Courtesy of The Disney/ABC Cable Group.

90 Courtesy of Warner Bros.

91 (top and centre) Courtesy of Peter Munk. (bottom) NSARM (vol. 2183, file 15).

92 DX.

93 Courtesy of Peter Munk.

94 DX.

95 From Clairtone brochure, 1966. DX.

Era Two

96 Courtesy of Jack Lane.

97 DX.

98-99 DX.

100 DX.

102 From Clairtone "Home Decorator Preview Kit." DX.

103 NSARM (vol. 2225, file 1). Photo by Peter Croydon.

104 Photo by Ezra Stroller.

105 From Clairtone annual report, 1967. DX.

106 From Clairtone annual report, 1965. DX.

108 (left and right) DX, Al Faux archive.

109 (top, right and left) NSARM (vol. 2170, file 9). (bottom) DX, Al Faux archive.

110-111 Collection of Linda Munk. Photo by George Whiteside/© DX.

112 DX (991.1). Gift of Frank Davies. Photo by George Whiteside/© DX.

113 Collection of Linda Munk. Photo by George Whiteside/© DX.

114 NSARM (vol. 2227, file 10).

115 DX (991.1). Gift of Frank Davies. Photo by George Whiteside/© DX.

116 Courtesy of Vladimir Jidkoff.

117 NSARM (vol. 2232, file 11).

118 Courtesy of Gloria Collinson.

119 (left) Courtesy of Jack Lane. (right) Courtesy of Peter Munk.

120-121 Courtesy of Canal+.

122 *Clairtone News*, March 1965. DX.

123 (top) DX. Photo by George Whiteside/© DX. (bottom) Collection of Daniel Leblanc and DC Hillier.

124 Courtesy of Gloria Collinson. Photo by Jack Marshall & Co.

125 (left) DX. (right) Courtesy of Gloria Collinson and Frank Davies.

126 From "Clearly Clairtone" brochure. Courtesy of Greg Yaneff.

127 Ibid.

128 Ibid.

Era Three

130 Courtesy of George Weil.

131 From Clairtone annual report, 1964. DX.

132-133 DX.

134 Courtesy of Gloria Collinson.

135 NSARM (vol. 2225, file 7). Photo by Everett Roseborough.

136 From Clairtone new product binder. DX.

137 DX (5116). Gift of Rodney Nash. Photo by George Whiteside/© DX.

138 From "Clearly Clairtone" brochure. Courtesy of Greg Yaneff.

140-141 DX.

142-143 Collection of the Fritz Spiess Archive, Media Commons Audio Visual Library, University of Toronto.

144 From *Better Homes and Gardens*, "Christmas Ideas" issue, 1967. Courtesy of *Better Homes and Gardens*.

145 DX.

146 From *Globe and Mail*, October 12, 1966.

147 (left) Courtesy of Gloria Collinson and Frank Davies. (right) DX.

148-149 From *Chatelaine*, January 1967. Photos by René Delbuguet. (far right) From *Chatelaine*, May 1967.

150-151 Collection of the Fritz Spiess Archive, Media Commons Audio Visual Library, University of Toronto.

152 (left) Courtesy of Tony Mann. (right) NSARM (vol. 2183, file 15).

153 DX.

154 (top and bottom right) Courtesy of Burton Kramer. (bottom left) DX, Burton Kramer archive.

155 DX, Burton Kramer archive. Photo by George Whiteside/© DX.

156 DX, Burton Kramer archive.

157 DX, Burton Kramer archive.

158 Sinclair Micromatic: Collection of Michael Chojnacki. Clairtone Mini Hi-fi: DX collection (5003). Gift of Arthur Deutsch. Photo by George Whiteside/© DX.

159 (left) Courtesy of Rodney Nash. (top and bottom right) DX.

160 DX.

161 Collection of Daniel Leblanc and DC Hillier. Photo by George Whiteside/© DX.

162 DX.

163 (top left and right) DX. (bottom) Courtesy of Gloria Collinson and Frank Davies.

164-165 Courtesy of *Canadian Interiors*.

166 From "Clearly Clairtone" brochure. Courtesy of Greg Yaneff.

172-173 DX (998.9). Gift of Peter Munk. Photo by Guy Lavigueur, Productions Punch Inc./© Canada Post Corp.

174-175 Collection of the Fritz Spiess Archive, Media Commons Audio Visual Library, University of Toronto.

176 DX (998.9). Gift of Peter Munk. Photo by George Whiteside/© DX.

177 DX (991.1). Gift of Frank Davies. Photo by George Whiteside/© DX.

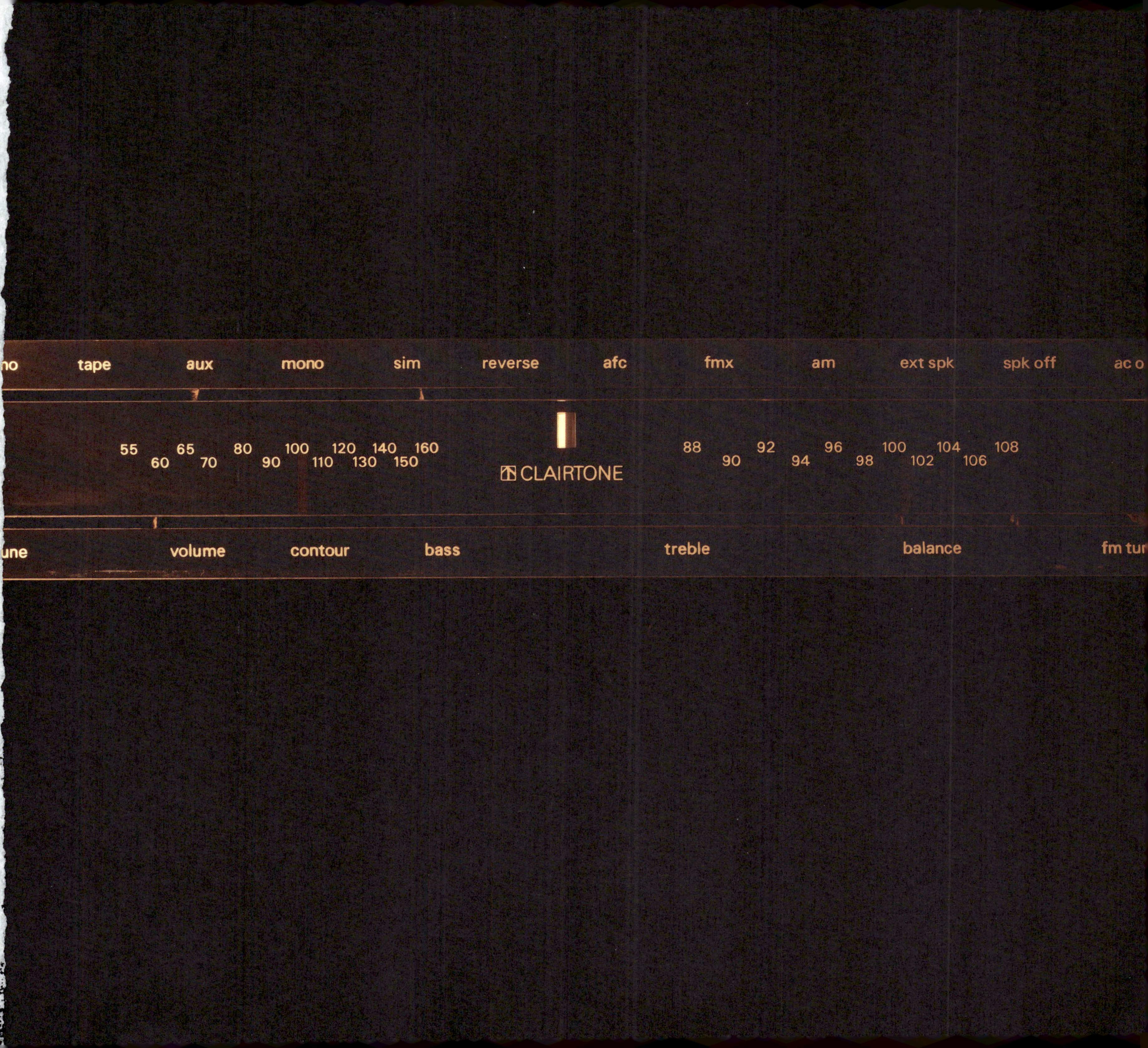
tape
aux
mono
sim
reverse
afc
fmx
am
ext spk
spk off
55 60 65 70 80 90 100 110 120 130 140 150 160
CLAIRTONE
88 90 92 94 96 98 100 102 104 106 108
volume
contour
bass
treble
balance